WHEN TIMBER STOOD TALL

BY JOSEPH H. PIERRE

Heavy stand of old growth yellow fir. Figure standing by a fir 98 inches in stump diameter. *Courtesy Oregon Forestry Department, Salem, Oregon.*

SUPERIOR PUBLISHING COMPANY

BOX 1710 • 708 SIXTH AVE. NO. • SEATTLE, WA 98111

Library of Congress Cataloging in Publication Data

Pierre, Joseph H
When timber stood tall.

Includes index.
1. Lumber trade—The West—History. 2. Lumbering—The West—History. i. Title.
HD9757.A4P53 338.4'7'6740978 79-21006
ISBN 0-87564-909-2

FIRST EDITION

Photographic reproduction by Artcraft Colorgraphics - Seattle, WA.
Layout and book design by Phyllis Berg
Jacket concept and design by Joe Salisbury

Printed and bound in the United States of America

DEDICATION

Dedicated to those men who, rightly or wrongly, assailed and took the most massive, magnificent resource above ground on the North American Continent—the beautiful, giant timber.

CONTENTS

INTRODUCTION

As a windjammer sailor becomes so attached to the life at sea that no matter how many times his ship fought for her life in typhoons that carried away the rigging, and sometimes dismasted and nearly foundered her, no matter how many times he had felt the cold hand of death very near, he still wanted no other life, because the glory of the sea was in his blood ...so it is with a man born and raised around the big timber. He may have labored to exhaustion, drenched with sweat in the summer. He may have fought the rigging frantically and doggedly in the bitter cold of the winter. He may have helped carry out friends who were struck down beside him by a flying choker or a broken snag, or who had been run over and crushed by a rolling log. He may even himself have been injured in the woods more than once, and missed death by only inches many times. None of these things, or all of them, will change him.

Like the sailor, the logger is the victim of a certain obsession, a fever, and he will never willingly turn his back on his way of life. It is in his blood—and when he grows older, if he does, his memories will not carry him back to love affairs, or picnics, or trips, or fishing, or hunting—half so often as they will carry him back to the sights and the sounds of the old logging woods. The raucous blast of the donkey whistle still rings clear and compelling in his nostalgic mind. The boom of log colliding with log or stump, moved by the irresistable force of the taut mainline is still there. The feel of the tough leather rigging gloves and the snugly fitting caulk shoes and the stiffness of the "tin" pants and the leaden, dead weight of the stiff wire rope of the thick choker. These things have been burned into his memory, never to leave until all else is gone from his mind. Logging camps were for men. Two things were uppermost in the minds of those who fought for the massive prizes nature had to offer—the big trees; *get logs,* and stay alive. Those who took their places in the woods in the old days of "highball" logging became adept at both, or they didn't last long.

Of the several varieties of timber common to the Pacific Northwest in the early 20th century, Douglas fir was the most plentiful. Other varieties were common to certain areas such as the large spruce found in Western Oregon and Washington, along with a considerable amount of hemlock, white fir, and cedar. East of the Cascades, pine was, and is, more commonly found than any other kind of timber.

It might be mentioned that various woods are more suitable for specific uses: Hemlock, for example, makes an excellent floor when cut with the grain vertical to the surface, and cedar, which has been so widely used for shingles and shakes. Such special uses of these and other varieties of timber are in addition to their general use for lumber. We are not so concerned here with the uses of the timber, however, as we are with the cutting, logging, transporting, and milling of it.

The men who worked in the woods differed from men in a hundred other occupations. They too were capable of expressing and appreciating subtle humor. They laughed with enthusiasm when something was really funny, but they had a language of their own, a Western dialect as unique as any in the land. They liked conversation when they had time for it. They liked good food and a good bed. The logging camps that had trouble keeping men usually failed to provide one or the other, or both, of these.

The way in which loggers differed from the average American was that they were tough, courageous, hard-bitten, rough men. Their work made them that way. A buttrigging with its dangling, whipping chokers travelling back into the woods from the donkey at sixty miles an hour was not something to encourage the presence of the cowardly or weak. When the whistle punk pulled the whistle and a turn of logs weighing ten tons began to move it took only a loose vine maple branch to trip a man attempting to get clear, and bring about sudden, horrible death. These men lived with that awareness every day. Each man of them had known good friends who had met their end in the woods. Each man had had his own close calls. None of them having worked in the woods for any considerable time, could have failed to see at least one dead or injured man on his way to town on a flatcar. So why did these men continue such a life of chance, of constant danger, of hard, brutal toil, wilderness and hardship?...Because they were loggers. That was the only reason they needed. They were not concerned that the building industry needed lumber, there was no thought given to the effects on the economy of a high production of lumber. They didn't give a damn about those things...they were loggers. If you offered one of them a life in the Virgin Islands, he would have asked "What kind of timber they got down there?"

The typical logger was a man who would joke with you, drink with you, or fight with you with equal enthusiasm. Unfortunately the joking and drinking were much to be preferred over the fighting. Loggers mostly subscribed to the old sailor's philosophy that if you got a man down you'd better mark him good or you would have to do it over again sooner or later. The caulked boots were good for putting one's brand on a man. A man kicked across the face with a caulked shoe was not likely to come back for a second treatment. However, these men were not abnormally mean; they were the product of the environment. Engaged in a struggle, a man was impelled to do what he knew would be done to him if he were the loser. The fighting, when it *did* occur in camp, usually happened in the bunkhouse, then mostly on weekends when the lack of something to do was pressing. When the men were working they were too hot and tired to fight, and were more apt to indulge themselves with a quick joke than anything else.

In the days of railroad logging, camps were big and far back in the mountains. The only contact with town was the log train or sometimes a speeder. Unless a man quit, he didn't get to town except on Saturday, and he had to catch the train or speeder back in, some time on Sunday. The few hours spent in town were usually enough to do away with any accumulated funds. Except during prohibition, the men frequented houses of prostitution and saloons. While prohibition was in force, they went to "bootleg joints" and "road houses". The latter often provided women and rooms in addition to "moonshine" whiskey, and the houses of prostitution, which flourished during prohibition, served moonshine whiskey along with their other inducements.

It was a futile existence to say the least. Some loggers went to town only once a month. In twenty-four hours they would spend all they had made in a month. Some went to town only twice a year. With these, their money might last almost a week. There were exceptions, of course; those who had too much sense to go that route, and those who had families to provide for. These latter usually went home to their little "stump ranches" or whatever they happened to have, and laid in some groceries and possibly some clothing for their families, before they went back to the grind of decimating America's most beautiful resource—its timber.

When horses and oxen were combined in team of four, horses were put in front because of their inclination to walk faster. Note concave tires on wheels, and wooden rails. *Sketch, Joseph H. Pierre, Jr., Monterey, Calif.*

CHAPTER 1
BULLS AND HORSES

My father, my uncles and some of my older cousins spent memorable years at "bull logging" (ground logging with oxen). The huge beasts may not have possessed strength superior to horses, and they were slow, but they would lean into a load without the jerky and unpredictable movements common to the efforts of many of the horses. The ox yoke fitted across the upper necks of the team and they did their pulling by dropping their heads and leaning into the load. A good-sized ox could weigh 1400 to 1800 pounds. After the log was yarded, by a team of six to eight oxen, bigger teams, pulling as many as three logs were used to transport the load to water or to rail over a skidroad, a system of cross-ties shaped with an adz to accommodate either a sled or "sniped" logs (logs with bluntly tapered ends).

For over forty years, in general terms from prior to 1870 until 1910, the bulk of the logging was done by the use of oxen, or "bulls" as they were always called and sometimes were. There was no specific point in time at which the use of oxen was discontinued and logging began to be done by horses, just as there was no sharp point in time when both of these were supplanted by the use of the donkey engine. Many of the old loggers laughed at the donkey engine when it first began to make appearance in the Northwest, saying they would stick with the bulls.

It might understandably be questioned how bulls could possibly transport enough logs to keep up with a productive logging operation. The answer is surprisingly simple—just have plenty of bulls. Most bull teams consisted of ten or more animals; 12 was not

Team of 12 bulls, Tillamook, *Courtesy of Lila Boge, Tillamook, Oregon.*

Tillamook Company. *Courtesy of Pioneer Museum.*

Ox teams, note ox goad in the hand of the bullwhacker. *Courtesy of Lila Boge, Tillamook, Oregon.*

uncommon. The teams passed each other going to and from the woods. The skidroads were often one or two miles long. If the distance logs must be hauled was excessive, the logs were loaded on heavy-duty wagons in the woods and were pulled by bulls or horses, or both, to either water or a railroad.

"Bullwhackers" (drivers) used no lines. Their bulls were managed with two tools: the ox goad (a long stick with a spike in the end) and a bull whip which could be made to pop like a gun close to the ox's head. Most bullwhackers did not actually strike the bulls with the whip. It wasn't necessary. With those two tools the driver could make his animals do anything he wanted, including turning the long team in a tight circle at the point where he must hook onto logs to be yarded. As with horses, most of the control was verbal. Many logging outfits had four to six dozen bulls coming and going all day long. On the skidroad, ten bulls could pull three good-sized logs chained together end-to-end by chains which had a specially designed hook on each end. These hooks were driven into the sides of the logs near the ends, and the chains were crossed between fore and after logs. This had two purposes: it helped to keep the hooks from pulling out, and it allowed for the logs to negotiate the gentle turns without lessening or increasing the strain on either chain.

Skidroads were built with the skids spaced to accommodate the length of the step of the average bull which always stepped over the skid, thus walking always on the ground. The center of each skid was hewn slightly concave to keep the logs always in the center of the skidroad. Teams of bulls returning from the dump without a load had to move to allow the load to pass, as it could not move over because of the hollow of the skids. It was necessary for the skids to be greased; therefore, it was the regular job of one man to travel the road constantly with a bucket of skid grease and to splash the heavy material, somewhat similar to crude oil, on every skid. This made the loads pull easily. It was also important that skidroads be laid out so that the load was always pulled on the level or slightly downhill. Animals could not pull much of a load uphill, even if the grade was slight.

Many loggers preferred the use of the mammoth sleds on which multiple logs could be loaded for the longer skidroad hauls. The runners of these sleds were made from oak trees and tapered gently at the toes. They slid nicely along the skidroad, with the aid of skid grease of course, and fewer oxen or horses could pull a bigger load.

The Eastern method was different from that used in the West where the cold of winter was not so intense. In the Eastern winters bobsleds were used. They were designed only for operation on ice. to insure constant, solid ice on the road, a water sled was kept in operation sprinkling water on the roadway so the bobsleds would not wear away the ice. This operation was on the ground, no skids were used.

This brings us to the problem of shoes for the oxen. Their hooves were tough but they were not equal to the continual punishment of their work; furthermore, the bare hooves did not provide the traction needed. shoeing an ox, however, proved quite different from shoeing a horse. While a horse

Heavy duty 8-wheel log wagon used with horses in early peninsula logging. *Sketch by Joseph H. Pierre, Jr., Monterey, Calif.*

Team of 12 oxen pulling 2 logs. *Courtesy of Capitol Museum, Olympia, Washington.*

would allow a blacksmith to pick up one of his feet and calmly stand on the other three while one was being shod, bulls would have no part of this business of messing with their feet.

The problem was solved by the construction of a log crib into which the ox was driven, and completely enclosed. Straps under his belly kept him from lying down; logs across his back prevented him from rearing up. Each leg was secured and each foot, as it was being shod, was tied up so the smith could apply the two half-moon shoes to each foot, as the ox has a cloven hoof.

A bull is nothing like a horse. Generally much less nervous and skittish than the horse, he is nevertheless somewhat more difficult to handle in some ways. An actual ox is really a steer, a bull that has been deprived of his natural privileges. Regular bulls were used along with the others. They were oxen by profession, not by reason of changes made in their physical attributes. These could be dangrous at times, though they rarely gave trouble to their regular handlers. The "bull cook" who fed and bedded the animals seldom had trouble with them either. He was considered their benefactor, and they looked to him for feed. (The term "bull cook" later came to mean a general roustabout for the camp cook.) Being a ruminant, the ox would spend hours each night munching his hay and grain.

One curious incident took place near Black Lake, Washington, in the old skidroad days. One of the oxen got onto the railroad track and was struck by a train and killed. The manager of the bull-logging operation asked the railroad company to pay the value of the ox. The request was granted without question. Two days later, it came to the attention of the railroad company that the bull-logging outfit had dressed the ox and fed him to the men in their cookhouse. The railroad then requested payment for 1600 pounds of beef, stating, and quite correctly so, that the animal had belonged to them from the moment they had paid for him. Unhappily, but wisely, the bull-logging outfit paid the price of beef for an animal fit only for work which was considerably more than they received. There were those who thought it served the outfit

Ox being shod in special crib that restrained his movement in any direction, each foot required two shoes, as hooves are cloven. *Sketch by Joseph H. Pierre, Jr., Monterey, Calif.*

right for serving an old bull to their men for beef. During the same period that oxen were used to log, horses were also used, but horses continued to be used long after there continued to be a plentiful supply of oxen broken for work.

A good team of horses, handled by an expert driver, could be something almost inspiring to watch. The driver did not use the reins (they were invariably called *lines*), he talked to his team. *Gee* meant go to the right, *haw* meant go to the left. *Get up* meant go and *whoa* meant stop. *Back* meant exactly that. All this was seasoned by a hundred expressions such as "get in there", "easy", and other expressions that the horses understood perfectly. It was *something* to see a real expert, such as my father, stand on a stump and talk his horses through an entanglement of brush and stumps while yarding out a big log to the point where he could jump on the log and ride it out, though he would generally walk behind most of the time rather than burden the already heavily loaded team with his added weight.

We have been talking about a team of no more than four horses yarding logs out of the underbrush, but hauling logs to the landing often involved the use of six or eight horses which could still be handled to a great extent by talking. Some drivers could turn eight horses into the landing and spot the log, or turn of logs, to within inches of precision, by spoken commands. The landing might be at a railroad or at water where the logs could continue to the mill. If the landing was at water, sometimes the water was not sufficient to sustain a year-round operation; in that case water would be stored behind a dam above the log dump, and the logs would go downstream only when the dam was opened for a "splash" which would create an effective flood of the stream, carrying the logs in one giant mass down the stream to the mill below where they would be held in a "boom" until one by one they had been pulled up into the mill for cutting into lumber. Work went on 12 hours a day, six days a week. In some camps the teamsters slept in the hay in the horse barn. Things changed as time went on, especially with the innovation of organized labor. Logging companies began to provide bigger and better bunk houses, and teamsters were allowed to bed there. The food served was better, and the work-day shortened, first to ten hours, then to nine, and finally to eight. It was a long time before the work-week was shortened to five days.

The story of the development of organized labor in the Northwest can be touched upon only briefly here. Many books have been devoted to the subject, and it concerns us only in that organized labor, the Industrial Workers of the World in particular, had such profound effect in Northwestern logging camps that it cannot be ignored. Rather than to introduce the subject repeatedly, in bits and pieces throughout our story, it seems wiser to project a bit in the chronoogical order of events to offer here a brief summary of the development of the I.W.W., the L.L.L.L., and the wartime *Spruce Production Division*. While the I.W.W. was an active, and radical, organization for more than 20 years, the years of crises were 1916-1919.

Quoting from Robert Tyler's *Rebels of the Woods*, "One protest movement, the I.W.W., appeared in the region during this era and for a brief period during the first world war grew to alarming proportions. I.W.W. members preached the most 'advanced' radicalism of the day, an Americanized

Splash dam opened to flood stream. Thurston County, Washington. *Courtesy Capitol Museum, Olympia, Washington.*

anarcho-syndicalism, as they closed in combat with the developing capitalism of the Pacific Northwest".

When the I.W.W. decided to go along with the proposal to strike by the A.F.L. Union, the rather ineffectual *International Union of Timber Workers,* Wobbly delegates in the mills and logging camps passed the word to their comrades, "The strike is on". With the *Timber Workers* joined with the I.W.W. (Wobblies) as co-belligerents, if not actual allies, reliable reference reports indicate that 90% of the mills and logging operations of the Pacific Northwest were closed down. The strike came July 16, 1917.

It was the policy of the I.W.W. to have as many "walking delegates" as possible, in as many camps as possible at all times. The men in those camps, subjected to 10 to 12 hours of work a day (even 16 in some cases), poor food consisting mostly of starches and grease, wooden bunks with no bedding or linen provided, but straw ticks populated with bed bugs, overcrowded unventilated bunkhouses, and no provisions either for bathing or laundry, proved receptive audiences to the radical leadings of the I.W.W. organizers.

The Wobblies experienced the flush and glory of power for a time, but bloody days were ahead. In Aberdeen, only four miles from Hoquiam where the I.W.W. had experienced one of their most spectacular victories, calling more than 3000 men off the job in one day, and being supported by a sympathy strike of the shipyard workers of Grays Harbor County, I.W.W. leaders, organizers, and "walking delegates" were herded into boxcars and transported out of the country never to be heard from again.

In Everett, Washington, sheriff's deputies, backed by the Commercial Club and other anti-union enthusiasts, met 41 Wobblies as they disembarked from a passenger boat and clubbed them with night-sticks and revolver butts. Loading them into trucks they took them to Beverly Park where they followed the preliminary beating with another, *quite* thorough.

Because of the vioilence, 300 Wobblies left Seattle on the *Verona* for Everett. Word got to Sheriff McRae that the Wobblies had taken ship in Seattle. He gathered a virtual army of deputies and "patriotic volunteers". Meeting the *Verona* at the dock, a withering fire was directed at the men on deck. Five Wobblies were known dead, after the *Verona* had backed violently away from the dock, breaking its bow line. Thirty-one were wounded, two of the sheriff's force were dead, and 19 were wounded.

The beginning of the end came for the I.W.W. after the 1919 Armistice Day riot in Centralia, Washington. The Legionnaire's platoons were leading the parade. They stopped in front of the I.W.W. hall to close ranks. One man broke ranks and dashed to the door of the hotel containing the I.W.W. hall. A storm of men followed. Firing began from across the street where the I.W.W. had moved for fear of the impending attack.

Men died that day, Legionnaires and I.W.W. Some who did not die were beaten and their teeth kicked out. During the night, the lights of Centralia mysteriously went out. One of the Wobblies, known to have shot a man, was taken from jail, taken in a car to a railroad trestle south of Centralia, and hanged. Not once but twice, the second time with a longer rope so he could be dropped farther. Most accounts assert that Everest, the Wobbly, was castrated with a razor blade on the way to the trestle. Cut down the next morning, the body lay in the shallow river below until evening, when it was removed and placed in the jail with the other Wobblies and left there for two days.

Coroner Livingstone, in whose car Everest had been transported, submitted this report: Everest had broken jail, had fled to the river, had hanged himself, had then climbed back up the rope to replace it with a longer one, had jumped again, had then shot himself several times, and finally had drowned in the river after cutting the rope.

My mother hated the I.W.W. with a passion. My father was noncommittal. Friends and relatives have been divided in their feelings. One thing all agreed on, the I.W.W. forced changes.

Two deterrents to the surge of power of the I.W.W. and other lumbering-logging unions, came into being during that era. One was the Loyal Legion of Loggers and Lumbermen (L.L.L.L.) The other was the commissioning of the Spruce Production Divsion. Both were engineered and headed by an ex-army officer, ex-prison warden Colonel Brice P. Disque. He began recruiting for the L.L.L.L. in 1917, and by 1918 the organization had thousands of members in the Northwest. Never intended to be a labor union as such, the L.L.L.L. virtually became one in spite of the wishes of Disque to the contrary.

Provided with 100 officers and 25,000 men to put into the camps, Disque proposed to force production

A camp of the Spruce Division in Western Washington. *Sketch by Joseph H. Pierre, Jr., Monterey, Calif.*

of the needed 10,000,000 feet of airplane spruce needed each year. The Spruce Production Division was created in 1917 at a time when annual production of Sitka Spruce was no more than 2,600,000. Although the army men were supposed only to aid in the production of spruce, it was noted with alarm by the loggers that soldiers were also put into camps that had no spruce timber to be logged.

Given power from Washington, (he had reenlisted), Disque ordered the eight-hour day, the provision of bedding and linen service in the camps, and the establishment of the first mandatory minimum wage, ranging from 40 cents per hour for common labor to as high as 90 cents for some skilled labor.

In the end, accedence to some of the basic demands of the I.W.W. accomplished what opposition had not. Men quit the virtual sit-down strike and the aircraft industry got its spruce.

Typical geared engine used in early railroad logging; the Tom Bordeaux (Climax). *Courtesy Custom Photo Service, Littlerock, Washington.*

CHAPTER 2
STEAM

Two major innovations came to the Northwest logging industry. One was the logging railroad with specially constructed geared engines for steep country. The other was the steam donkey.

Where railroads had been used to move logs to distant points, after the "bulls" had logged them to "shortline" railroads, or mainline spurs, it gradually dawned upon logging operators of the 1870's and early 1880's that timbered hills could be penetrated deeply and efficiently only by rail. Too much of the timber lay a long way from water, a standard means of transporting logs, and a great deal too far from practical mill sites to transport them by bulls or horses. Manufacturers of rolling stock responded to these needs by producing engines suitable for the gruelling work, and railroad cars equipped with log bunks for transporting the timber.

The donkey engine, an adapation of the steam-powered capstan used on ships, and called by the same name, was first designed by John Dolbeer for use in moving logs in Humboldt County, California. Heavy hemp rope was used to pull the logs. The donkey had no drum. It was the work of one man to take two or three turns around the upright capstan, holding the slack and coiling the rope on the ground as the log was pulled in. Logs were at first only yarded 150 to 200 feet to the skidroad, and slow, plodding, powerful teams of as many as 12 oxen took them from there to the water, or rail spur.

But donkeys came into their own. Within a few years no less than a dozen companies were manufacturing donkeys. One company alone manufactured 26 different types of steam donkey. By 1913, the Willamette Iron Works had the capability of producing one complete, finished donkey a day. All were built to fill orders and were shipped to the purchasers immediately upon completion. A rather graphic picture of the possibilities of the donkey engine is seen in the true story of Anthony Dwyer who bought a big donkey and a 42-ton locomotive from the Willamette Iron Works. Barging the two pieces of equipment to Kalama, Washington, he loaded them on sleds and used the power of the donkey to pull the two more than ten miles, to the Coweman Valley over a 1500-foot elevation on the way.

With such power available, hemp rope was obsolete for use as yarding line. Steel manufacturers began producing "wire rope", also called cable, but in the woods simply called *line*. Strong steel pulleys, called *blocks*, were developed to lead the lines where desired. The line horse, used to pull the old hemp rope out from the donkey, could not be used to

Two donkeys and locomotive with loads at landing, Mason County Logging Co. *Courtesy of Capitol Museum, Olympia, Washington.*

With railroad yet to be finished, a "new tree" has been rigged. Donkey is ready to log. *Courtesy University of Washington Library, Kinsey Collection.*

pull heavy steel cable hundreds of feet from the donkey, nor was this method efficient in view of the speed donkey drums were capable of moving things. The sweating, swearing, rig-up crew of six to eight men went out first, dragging the three-eights-inch guinea line straight to the tail block, hung on a tree or stump at the back of the cut timber, from there to the corner block, out of the way of the logging, and back to the donkey. There it was secured to the haulback line and the guinea line drum was engaged. From there on it was simple to finish the ground rigging. The donkey puncher pulled the haulback out through the blocks and back. It was fastened to the mainline and its buttrigging and the puncher pulled on the haulback until the mainline was within reach of logs.

Thus rigged, it was time for the choker-setters to go to work. The "riggin' slinger" (lead man of the choker-setters) used the full length of his chokers, plus all the slack he and his men could pull in the mainline to reach every log possible before stopping operations to move the tail-block for a new swath. Logs partially embedded in the ground had to have a "choker hole" either dug or blasted to get the nubbin end of the choker under the log so it could be pulled through and buttoned into the bell. The early chokers, however, called "Peter hooks", were worse because the huge flat hook itself had to be poked through the choker hole. A half-stick of blasting powder was frequently used to blow out the hole.

The powder monkey was not called upon for the simple, small shot needed for a choker hole. The riggin' slinger did it. The powder monkey had enough to do without that. Bulls could go around stumps. The donkey could pull only in a straight line. When the powder monkey found that turn after turn of logs hung up on the same stump, he removed it. He normally blew several small stumps for each change of the tail block. This was the smaller part of his job. He spent more time blasting stumps and rock out of the right-of-way for extensions of railroad. A powder house, removed from other buildings, always contained plenty of powder, fuse, and blasting caps. When the quitting whistle blew, the powder monkey might have several shots charged, waiting for the crew to leave the area before he set them off. It didn't pay for him to lose track of what he had done. One powder monkey did, having forgotten that he had heavily charged a stump the day before. He rammed his bar under it, crimped a percussion cap onto a length of fuse with his teeth, the cap containing enough high explosive to shatter his head, inserted the cap into a half-stick of powder and poked it into the hole. His intention was to blow a pocket under the stump to hold a full charge. He lit the fuse and jumped around corner of the stump to wait for the "pocket shot" to go. With now more than 20 sticks of powder under the big stump, when the explosion did come it blew the stump up and out, and the powder monkey with it. He hit the ground a hundred feet from the stump, unconscious but not dead. He had many broken bones, damaged ears and eyes, and blood poured from his nose. He was alive only because he was pressed against the stump and was thrown *with* it, not struck *by* it.

Because of the continual problem of hang-ups on stumps, loggers experimented with the use of *lift* in yarding logs. At first, a "high lead" block was hung only a little way up a tree. This proved such an advantage that intelligent loggers wanted to pull from as high as possible, to practically eliminate the problem of hang-ups.

A whole new, dangerous, spectacular job came into being, high rigging! Pole climber's spurs were no good. Fir bark was often eight inches thick. A pole climber's belt was not adequate. The highclimber had to have special spurs, a special belt, a specially constructed rope with a steel core so if he hit it by accident when chopping limbs he would not come crashing to the ground, a special light ax and a short saw, both tethered to his belt to free his hands for climbing.

A proud, daring skill was developed that lives on today, to ascend, top and descend a spar tree in the shortest possible time. The high climbers were a fearless, and very high-paid special kind of loggers. Having climbed, limbed, and topped a spar tree, the high climber almost invariably sat on its still swaying top for a rest before he came down for the first of many items he had to take up and hang in the tree's top. One I knew, George Dunn, often stood on his head on the top of the tree as it still wobbled from the kick of the falling top. Was he an exhibitionist, a show-off? I don't know. If he was, perhaps so are the circus high-wire performers. I know the men watching him from the ground loved it.

The capabilities of the donkey engine were no less than fantastic. A competent puncher could move his donkey from one setting to another, with the aid of the crew who secured the mainline to one stump or tree after another as needed, in a very short time. He could get logs out of tough situations by braking the haulback and pulling on the mainline, or vice versa. The technique was called "tightlining". Because the bull block (high lead block) was high in the tree, the tightlining method would lift the troublesome log straight up. If the puncher happened to be running a swing donkey or skyline donkey, he could move 20

Climber topping tree, he has finished the cut and top is falling.
Courtesy Capitol Museum, Olympia, Washington.

Spar tree and donkey; tree just topped. Climber sitting on top while tree is still in motion. *Courtesy Custom Photo Service,*

Donkey and crew.

tons of timber from the tail tree to the head tree so fast that with the lift of the skyline carriage the turn of logs was in the air half the time. If his donkey was a roader, one that dragged logs over a continuous cradle of embedded logs, he could pull 20,000 feet of timber over a mile of such pole road. The road might be located at a rail spur, or at water. The expediency of using a pole road could be that its use would be cheaper for the amount of timber available to it than it would cost to build a railroad, or there might be no railroad near and the pole road's terminal was a river or a bay.

Often the pole road was so steep that the log needed only a start at the top to slide at ever increasing speed until it struck the water at a hundred miles an hour, scorched from the friction of its blazing speed on unlubricated poles. These were called chutes.

Those old wood-burners kept as many as four men very busy. The puncher had his work that required his undivided attention. The fireman crammed the firebox with wood every time the donkey was not pulling. He constantly watched the water glass and injected water whenever it dropped a bit in the glass. On a big donkey, two buckers were wasting no time cutting blocks off wood logs and the splitter quickly reduced them to pieces the fireman could use to advantage to make steam. The number of men needed to keep such a donkey operating was one of the drawbacks of a wood burner. But that was not the only disadvantage. When the donkey was pulling hard, the forced draft expelled burning coals as big as a man's thumb from the stack, creating a constant fire hazard in summer months. A third, and increasingly important factor, that eventually brought an end to the use of wood for fuel was the value of the raw product itself. When donkeys were converted to burn fuel oil, only two men were needed, the puncher and the fireman. When the steam donkeys began to disappear altogether to be replaced by internal combustion diesels, only the engineer was needed. Admittedly, the old feeling of unlimited power was no longer there, it took twice as long to get a turn of logs to the landing, but the cost of three or four men was saved, valuable timber could be sold, instead of being burned, and the danger of fire was practically eliminated. Yes, that was another thing, in the days of wood burners, a fire watchman had to be kept at each logging site each night, in case a smouldering spark might be in a bit of moss, bark, or dry wood, and flare up in the night.

CHAPTER 3
TIMBER COUNTRY

If you took the road from Rochester, Washington to Independence Creek you would pass through what used to be the town of Independence, once the headquarters of the Independence Logging and Lumber Company. Following the old road up the valley of Independence Creek about six miles you would come to a fork in the road called the "boot jack" because the adjoining roads left an unsurfaced triangle in the middle of the fork. The road to the left crossed a flat and wound over the hills to Lincoln Creek.

About half a mile up that road, nestled by the foot of the hill, there was a log house built by my father on his homestead. It was in that house, December 8, 1906, that I was born. It was a wild country in those days. Deer were so plentiful that many cougars roamed about the hills because cougars live on deer.

On the afternoon of the eighth of December my mother was in labor. Women in those days were not often attended by doctors. Pioneer women helped each other in childbirth. My father was at home at the time, and my mother had sent the other children to the homestead of another pioneer family with the instruction to tell Mrs. Sturdevant that she would like her to come over. The kids started out walking, and arrived at the Sturdevant place eventually. The distance between the two places was more than a mile. The Sturdevant children met my brothers and sisters as they approached, and the whole bunch began to play games.

Some time had gone by before it began to dawn on Mrs. Sturdevant that there was something strange about all my brothers and sisters being allowed to come to visit at once, particularly since the distance was what it was and some of the children were pretty small for such a walk. She came outside and called to the children, asking how my mother was. She knew the time of delivery was near. One of the kids said she was fine and wanted her to come over.

Suddenly the truth came to Mrs. Sturdevant. She knew now why all the kids had been sent away from the Pier homestead. She hurried to get ready. One of the Sturdevant men hitched up a team, and Mrs. Sturdevant soon arrived at my father's homestead, too late. The stork had beaten her there and my father had learned about delivering babies. He followed my mother's instructions carefully and I, the eighth and last child of a western pioneer family, was born. Perhaps it was the smell of blood; as soon as it was dark a cougar began to scream. He came right up to the house and paced up and down on a log that lay along the windows on one side.

Born in the backwoods, near one of the largest logging operations in that part of the country at that time, I seemed destined to live around logging camps and lumber mills most of my life. In later years, many times I have walked the lonely country roads at night and heard the scream of the hungry cougar. Sometimes I would see him cross the road with stealthy swiftness and sometimes he would be in a tree or standing on a log by a trail.

One of the loggers, walking back to one of the Independence logging camps by moonlight, thought he heard a noise behind him. He was sure it was a cougar. He quickened his gait, only to find that the sound speeded in tempo. Breaking into a run, the man covered ground at a terrific pace only to discover, just as he entered camp exhausted, that the sound was the friction between the two legs of his tin pants.

So wild and untouched was the country where I was born that from the Chehalis River and Grays Harbor on the north to Willapa Harbor on the south, and from the Pacific Ocean on the west to beyond the Cascade Range of mountains on the east, stood an unbelievable splendor of old growth fir timber, still virtually untrammelled in the early 20th century. Not that the same quality of timber could not be found

Independence Logging Company's little road engine.

also to the south and to the north of that area, but so vast and so wild was that country, and so void of roads, that decades went by before the various logging operations, working on that huge forest from three sides, had made serious inroads on the supply of timber.

Railroads were already in use and the abandoned railroad grades still remain to be used by deer and elk hunters and by the forestry for fire roads. In those days there were no chain saws. The tools used to fall timber consisted of a falling saw, sometimes as long as 14 feet when the stand of timber was generally big, one eight-pound sledge, at least two falling wedges, an oil bottle for oiling the saw, two springboards to elevate the fallers above the undesirable swelled butt of the big trees, and two falling axes, often filed to a chisel bit to facilitate chopping the undercut. Falling axes had a long narrow bit and were double-bitted. Fallers wore stagged-off paraffin pants called "tin" pants because of their stiffness and resistance to snags, being made of exceedingly strong material and dipped in hot paraffin for waterproofing; and caulked shoes, which were invariably called "cork" shoes by loggers. The pants were never held up by a belt; heavy elastic suspenders were always used. The logger might wear a shirt, usually the time honored *hickory* shirt, and he might wear a waterproof coat and hat, but when he was working he was always stripped down to the traditional gray woolen underwear above the waist.

In big timber, it might easily take more than an hour to prepare one of the giant firs for falling.

Author's father, Dave Pier, on ground with small switch engine. Independence Logging Company.

Independence Logging Company's #1 3-truck shay. *Courtesy Pete Replinger.*

Author's father, Dave Pier, 3rd from left at Independence Logging Company, Mason County, Washington.

Swamping out had to be done to get rid of brush and debris around the tree, then notches had to be cut into the tree for the "bit" of the springboard. Sometimes several notches had to be staggered up the trunk which could be used as steps down from the spring board set in the highest notch. After the springboards were set so that the fallers were above the swelled butt, the undercut was sawn, then chopped out at an exact right angle to the spot where the tree would fall. So skilled were those old-time fallers with a falling ax that the chopped face of the undercut was nearly as smooth as the saw cut. Chips nearly as big as dinner plates would fly to accumulate at the base of the stump. With inhumanly powerful blows the two fallers swung their axes intermittently to avoid hitting the other's ax as the center of the cut was chopped. It was the boast of many a faller that he could place a stake in the ground, 150 feet from the stump, and drive it with the tree when it fell. This was no idle boast, the degree of accuracy was that great more often than not. After the undercut was put in with all that precision, the pair of fallers would bounce up and down on their spring boards, kicking them around toward the back of the tree each time their weight was off, to allow the men to turn and begin the back cut. All this was done from six to twenty feet off the ground. The back cut was then put in, using wedges to throw the tree, if that was necessary, holding precisely the right amount of wood on each side so the tree would not deviate from its planned position when it fell; and the lie of the tree was important, trees falling across stumps, logs, rocks or even solid mounds of earth could break and spoil a lot of good timber. In the big redwoods of California, beds of boughs and dirt are prepared to save the big and very breakable trees from damage.

The back-cut is put in at a higher level than the undercut. This is one of the means of insuring that the tree will go where the fallers want it, and will not split or kick back over the stump. when the back-cut is nearing completion the fallers will call back and forth how much wood is still holding on each side, working frantically when the tree first begins to lean, to reduce the holding wood as much as possible before the tree goes, but *without taking too much.* Bad things can happen if a faller cuts so deep that the remaining wood is not sufficient to hold up the tree. The tree can settle on the saw; it can twist off the remaining wood and go far enough off its intended direction of fall to hit some obstacle and break. Worse, since its action is no longer in control, it might fall sidewise, kick off the stump and kill a faller or hit someone else in the vicinity.

It is not uncommon for a big old-growth Douglas fir to reach a height of more than 250 feet. In a logging operation it is possible that a bucker, a scaler, the bullbuck, or even some of the rigging crew might be within the reach of the top of such a tree, especially if the tree does not go where it is supposed to go.

Beginning a couple of minutes before the tree will drop, the head faller will begin to yell, "Timber!" He does this so that anyone within 300 feet can get clear. This call is usually accompanied by a warning that is a little more specific, such as "Down the hill!" or "Up the hill!"

Some of the timber was so big the sides of the trees had to be chopped away to allow for the play of the saw. To me, even after many years of observing that particular thing, it is something awe-inspiring to watch a giant eight-feet through gradually begin to tip, going faster and faster as gravity has it captive, until as it crashes to the ground its tip has reached a speed of at least a hundred miles an hour, and the force of its fall can be felt in the ground for two hundred yards. It saddens one a little, too, to think of the hundreds of years that regal tree spent growing to that size, only to be cut down, bucked into logs and hauled off to some mill where it is cut into lumber.

Next in this process came the buckers. They measured off the tree, after they had limbed it, marked it for cutting into the most suitable lengths, and began cutting it at those marks. A bucker's tools consisted of a bucking saw, not so narrow and thin at the back as those used for falling, an undercutter (a tool with a roller upon which the back of the saw could ride and cut upward from the bottom of the log), a wedge or two, and a swamping ax which was also used for limbing. The bucker also had the responsibility of saving as much good timber as he could. It was also to his advantage to do so as both he and the fallers were "bushel workers", that is to say they were paid so much per thousand board feet for all they cut. Unlike the fallers, the buckers did not work in pairs. Each man was on his own. His work was not without hazards. If he got foolish and tried bucking a tree from the downhill side because it gave him a better place to stand, he might find that he had paid too high a price for that convenience when the log dropped, rolled down the hill and over the top of him. If he was working too close to the fallers, he might also catch a "widow-maker" (flying limb), or get hit by a snag that fell when struck by a falling tree. Bushelers could make better than average wages if they knew their business and were willing to work hard.

That is how it was at the Independence Logging Company in 1906, the year I was born, and for many years afterward. Railroads were the only possible way to reach the timber and get it to market. The locomotives were small and slow, but they did the job. The donkey engines were not as big or as powerful as some that came later, but they beat bull logging. My father worked for Independence, sometimes in the woods, sometimes on a donkey and sometimes running locomotive. My mother ran the cookhouse, a way up the valley from our homestead. Nobody then thought the timber would ever run out. You could stand on the top of the highest hill you could find, and see nothing but timber to the west, the southwest, and the northwest. It was still that way years later when I was old enough to appreciate it.

My first memory of timber and logging, however, was not in the giant timber. Some mill operators were already working in smaller timber. It was ideal for ties for the ever increasing new railroads that were being pushed into the big timber, and for the mainlines which always required supplies of ties for maintenance. My father took the contract to log for one such tie mill in 1911. It was located near the edge of the prairie about five miles southwest of Tenino, Washington. Here, too, my mother ran the cookhouse. I was nearly five then, and my brothers would sometimes take me up into the woods where I could watch the teams pulling out what looked to me like big logs. From a safe place, I could also watch the fallers and buckers. Then walking back we would follow a team with its log, and, walking barefoot, I can still remember the feel of the cool, mastic mud smoothed and shaped by the dragging logs.

It was there, too, that I saw my first truck. Always before teams had been used to haul the ties from the mill to the railroad. Now there was this new thing called a truck, said to be much faster. The truck came to the little mill and loaded, then proceeded to get stuck in the prairie before it got a hundred yards.

Loading with crotchline and spreader bar. Loader near tracks, yarder on right. *Courtesy William D. Jones Historical Collection, Aberdeen, Washington.*

Teams were brought down to pull the truck out, and the ties were again transported to the railroad by teams.

Because of the portable nature of the tie mill, and the temporary nature of the camp, no buildings were built. The bunkhouses were tents. Our house was a tent. The cookhouse was a tent. Except for the mill, the only thing around there that was made of wood was the outhouse. I'm sure there was more than one, but I only remember the one across from our tent where I would be escorted after dark because of the cougars. There were bears, too. One was attracted to the storage tent by the smell of dried apples which my mother used to make pies. He got into the barrel of dried apples and soon devoured them all. He was forthwith gripped by a prodigious thirst, and went straight to the creek where he drank and drank. The dried apples swelled and he drank some more. All this was reconstructed later by what followed. The night air was rent by a series of most plaintive moans interspersed with short howls. My father and the others went out into the night to investigate. Not far from the creek was the bear, his skin was so tight he couldn't walk, and his legs wouldn't even stick straight down. There was only one thing to do, that was to put the bear out of his misery.

The *bullbuck*, or foreman, of cutting timber was a good-natured Hungarian. He would barge into the cook tent looking for doughnuts, announcing, that he was a hungry Hungarian. He proved to be not quite so good-natured when a fight broke out in his bunk tent. His English was not too good, so we never did get the details of the fracas. It was Sunday and the men were all in their tents when the fight started. The uproar could be heard all over camp, since the walls were only canvas. Suddenly the door to the tent burst open and the hungry Hungarian (I never knew his name) threw a man out of the tent bodily, bellowing instructions as he did so to go to Centralia, go to Chehalis, go to Tenino, he didn't care "w'ere to hal you go, geet! Savey?"

The same Hungarian had little regard for money and he liked kids. I was pulling a pair of bottles harnessed with string to represent horses and hitched to a round chunk of limb which represented a log. I suspect that one of my brothers arranged the outfit for me. Anyway, the Hungarian saw me dragging the assembly near the bunk tent where he lived. He called out to me, "Ay keed, 'ow much you tek for dat team?" Surprised and confused, and knowing nothing whatsoever about money, I tried to think of some amount of money that I had heard of. I seemed to remember very little, but finally I answered six bits. I had heard my father use that term. "Awkeh," the Hungarian said, "I geev you seex beets." And he sure did. I proceeded to drive off with both the team and the "seex beets". Whereupon the Hungarian doubled over with laughter and nearly fell off the porch. Seventy five cents in those days was half a day's wages.

A stand of old growth. *Courtesy of Joe Bell, Tillamook, Oregon.*

CHAPTER 4
THE HIGHBALL DAYS

High lead logging had come into use early in the 20th century. It did not happen uniformly through the Northwest. While the high lead was being used successfully in some areas, and steam ground logging in other parts of the country, logging with horses continued in other places and bull logging continued in still others. One of the first parts of the Northwest to adapt to the use of steam donkeys was the Columbia River. Another area that came to the use of steam early was Puget Sound, around Seattle, although the first years that logs came into the Henry Yesler Mill in Seattle they came by bull teams down the skidroad which later became a street, Yesler Way. It was in Seattle that the term *skidroad* was used to refer to the part of town which was frequented by worn-out loggers, "winos" and bums. Later the term was corrupted into *skid row* but it was originally the *skid road*.

Highball logging was a study in efficiency, an efficiency that paid off only in the quantity of timber moved. In terms of human life, it was costly indeed. In a "highball" outfit, the choker setters did not stand clear as the buttrigging with its dangling chokers came to a stop in the woods. They were there when the chokers arrived, still whipping, to uncouple them from the buttrigging and hook up the chokers they had just finished attaching to several logs. Then as soon as the chokers were coupled to the buttrigging, the signal was given the whistle punk, and relayed that same instant by a single pull on the whistle wire. The donkey puncher quickly laid the power to the mainline and the turn of logs began to move. The men had to get clear during the time lapse, and a short one, while signals were being relayed and the donkey put in motion. The chaser (the man at the spar tree who unhooked the turn) jumped on the logs before they stopped moving and began to wrestle the chokers free. They had to be pulled out from under the logs by the donkey, and it was then that the logs often moved and a chaser had to watch out so he would not get caught by one of them. Two words were the order of the day, *get logs*. It was years before much thought was given to safety. Between 1910 and 1920 the Federal Government got into the picture and ordered several highball outfits to shut down temporarily as a penalty for fatalities that in some cases ran as high as an average of one man killed per day of logging.

One operator was finally ordered out of his own woods and forbidden to take any further active part in the logging operation. He was known far and wide for his insistence on *getting logs*. It has been told that when it was reported to him that a man had been killed, he would say to put him behind a stump in the shade, there might be another one before the day was over. Another story told about this man was that he walked into a pool hall stiff-legged as a banty rooster, and demanded to know if there were any loggers in the place. One man responded he guessed he was. When asked if he could get logs, he declared that he could. To the question, how fast, the man wanted to know if he was expected to fly. The operator asked if he could do that and the logger replied he could as long as he could light long enough to eat his lunch.

Such outfits were always in the market for loggers. Most men didn't care to work where there was no concern shown for the men, and would only do so if there was no work to be had any place else. After the high lead had been in use a few years, its potential was fully realized. The principle of lift was carried one more step. The skyline came into use. This greatly enlarged the scope of logging. The railroad only needed to reach the skyline donkey. The skyline could reach out 2500 feet or more to the tail tree. The skyline was pulled taut from one tree to the other, and secured to stumps, as were the guy lines from each tree. A carriage was mounted on the skyline. It had sheaves (or wheels) that rolled on the line and carried the mainline and its buttrigging back to the woods by the pull of the haulback.

From the "tail tree" (or yarder tree) the mainline of the yarder, its lead block also hung high in the tree and could reach beyond that point another 1000 feet or more. The effective result was that to one railroad spur, logs could be brought in from 3000 feet on each side, or a total swath on both sides of the railroad of a mile in width. This was logging on a big scale. The skyline made it possible to log over any kind of ground, even across a canyon.

New problems came with the improved methods. It took an unusually capable logger to "chase landing" at the tail tree. Here the chaser not only had to uncouple the chokers from the buttrigging of the yarder, he had to couple them to the rigging of the skyline donkey. Because of the great distance between the head tree at the railroad and the tail tree where the logs were switched from the yarder to the swing donkey, it became difficult to stretch the wire for the whistle, and keep it at the right tension. Electrically activated whistles seemed to be the answer. The electric wire could be strung out in any fashion most convenient.

In one camp, the whistle punk had been having difficulties with his whistle; sometimes it would fail to work. The chaser at the tail tree in this case was a young man who had been pushed forward a little too fast. He hadn't learned all the tricks of staying alive. Just as he finished coupling the turn of logs to the skyline rig, the whistle punk was trying to find out what was the matter with his whistle. Contact was made accidently, and the swing puncher up the hill heard his signal, popped the throttle and began the pull. The chaser got excited; instead of leaping to the side, he began to run toward the railroad, directly in the path of the turn of logs. He probably would have realized his mistake in a couple of seconds and changed his course, but he didn't have time. He was running in the stiff mud and in nothing flat that turn of logs caught him and ran over him. The crew of the yarder, and the whistle punk for the swing donkey saw the accident. The chaser had disappeared. They found him pressed down into the mud, and pulled him out. He was not dead. The yarder whistled the news that there had been an accident. The big two-speed at the railroad picked up on it and whistled for the locomotive. I was working at the head tree at the time. I'll never forget the way that man looked when they carried him up the hill to the railroad. He was so plastered with mud it was hard to believe I was looking at a human being.

Before long the old Heisler locomotive came roaring up the track full speed toward the landing. The victim of the careless accident was stretched out on blankets on a flatcar for the long, tortuous trip to the hospital. This involved the four-mile ride down to the camp, the transfer to the incline which went down the mountain half-a-mile at nearly forty percent grade. Engines could not be used there, logs were let down three cars at a time by a monster snubber with water-cooled brakes and a half-block purchase achieved by the bight of the snubber's mainline passing through a block car with three massive sheaves. At the bottom the injured man was transferred to the speeder and taken to headquarters, about 15 miles, where there was an ambulance. From there it was another 17 or 18 miles into town to the hospital.

I never saw the young chaser again, but heard later that he was recovering. He had many broken bones and other problems, but he survived. The yielding mud had saved his life.

Killed in the woods! How often we used to hear the words. My oldest brother, my uncle, my niece's husband, my closest friend, and many other friends and relatives were killed in the woods. For a period of a few years in the Northwest, for the number of people involved, the loss of life ran higher in the logging industry than it has in a number of wars. The accident rate is high enough, by the very nature of the work, even when all reasonable safety precautions are practiced. During that period when many logging operators showed little concern for human life, but a greed for logs, the death rate was three times as high as it was later on, when state and federal governments acted to enforce safe logging practices.

One instance is typical of the unpredictable accident that can happen even when no one has been careless, and no one rushing the work. I was working at one of the Mason County Timber Company's camps at the time, a camp called Gibson Creek. I was working at the swing tree. One thing was less than desirable, we were logging downhill. As mentioned before, logging uphill prevented most serious hangups, but in this instance to pull logs uphill would have involved the building of miles of railroad to reach a section or two of timber, so we were pulling it from 1000 feet back to the top of a bluff, then another 1000 feet down to the railroad. There were frequent hangups on top of the bluff because even our 200-foot tree did not reach up high enough to give any lift to the logs until they came down the bluff.

A turn of logs had hung up on a stump on top. The puncher of the swing donkey would reef on the mainline until the donkey stalled, drop some slack and then reef again. He usually got the logs loose that way. He was looking intently up the hill where the turn was hung on the stump. After the third or fourth strain the spar tree was wobbling violently in spite of its system of guy lines; some of us saw it but the puncher didn't. About the fifth time he threw the power to the main line, the spar tree broke in two in the middle. A dozen men were suddenly in peril as the whole

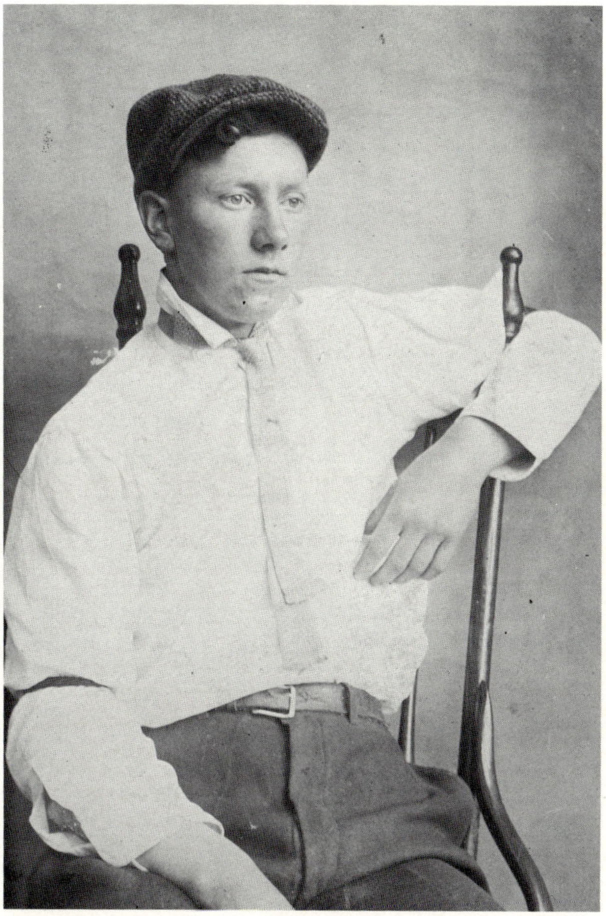

Oldest brother of the author killed while logging near Montesano, Washington.

umbrella of spar tree guy lines, mainline, haulback, bull block and the upper half of the spar tree came crashing to the ground. The puncher saw the skyline drop and he dove off the donkey. The chaser tried to dive under a log where there was only room enough for his head. He looked like the proverbial ostrich with his head in the sand, his knees were drawn up so that his posterior stuck up. Some ran, some didn't do anything because they didn't know which way to run. In seconds is was all over. The stack and roof were torn off the donkey. The rigging was in shambles. It took days to raise and rig a substitute tree so we could begin again, but though there were a dozen men under that tree with its total of eight guy lines, one skyline, one mainline and one haulback, plus two blocks (steel pulleys) the biggest of which weighed a ton, not a man got a scratch except the donkey puncher who skinned his knee diving off the donkey.

Raising a tree for a spar tree.

Loading donkey behind spar tree, tong line blocks away from tree half-way up. Yarder is out of sight, right of picture, turn of logs arriving at left. Mason County Logging Company. *Courtesy Capitol Museum, Olympia, Washington.*

CHAPTER 5
LIFE IN THE CAMPS

After the first two or three decades of logging in the Northwest, camps began to take on a uniformity. Not that conditions were uniformly of the same quality all over the country, that was not the case. There were camps that had trouble keeping men because of poor bunk houses, poor food, no showers, or the presence of bedbugs, while other outfits provided good beds, good food, and even in some cases reading and recreation rooms. But comparing the average camp with others of the same type, one would find a common pattern.

Some companies liked their camps on wheels. A bunkhouse would be left permanently on the flatcar that brought it there, and entrance to it involved climbing several steps. Other companies would slide the buildings off the cars on skids, so the cars could make the company money hauling logs. The row of bunk houses would be lined up along the railroad tracks, or a spur off them, with the cookhouse located centrally in the camp. Nearby would be the camp store, which carried most of the things a logger might need, from caulk shoes and tin pants to smoking and chewing tobacco, and snoose (snuff).

A typical bunkhouse would accommodate 16 men, eight in the lower bunks along the sides and eight in the upper bunks. The door was in the center on one side, and opposite it was the stove. Stoves were often built in the blacksmith shop from oil drums. They would hold plenty of wood and make lots of heat. This was important during the cold seasons when men would come in from the woods miserable from the wet or cold, or both. It was one of the duties of the bull cook to keep plenty of wood at each bunk house, and see that the fire was going when the men came in. Camps that were located near the center of 40 or 50 sections of good timber could be a little less temporary in nature. Some of the buildings, such as the cookhouse, were often quite roomy, some were as large as 30 by 120 feet. A camp running two or three sides (separate complete logging operations) might have to house and feed from 125 to 150 men. There would be a cook, a second cook, a dishwasher, and four flunkies (waiters), besides at least half the time, the bull cook handled the kitchen stores, fuel, and other things necessary to the operation of the cookhouse. The cook had to be both chef and baker. He planned the menu and ordered food supplies. When it was time to call the men to a meal, one of the flunkies went to the porch where a large steel triangle hung from a chain. There was a steel bar to thresh around the inside of the triangle. The resulting noise could be heard throughout the camp and far beyond.

In the evening, tired as the men were, they had energy enough left for some cards, reading or just shooting the breeze. Some bizarre and wild stories were to be heard after supper when the Swedes were munching on hardtack which the cookhouse always provided liberally, and others chewed tobacco and spat at the sand-filled box the stove stood in, and others were contentedly enjoying the numbing effects of a facefull of snoose.

It was on such an evening that Slim (the only name I ever knew for him) was holding forth in a voice loud enough to make it impractical for anyone to try to read. Slim was the puncher of one of the yarders. He enjoyed the glory of his job and loved to expound to the men who he thought knew less about donkey engines than he did. Dave, lying on his upper bunk, had been trying to read. Slim was saying he didn't know why they didn't build the drums on donkeys

Bunkhouse, loggers relaxing. Note ancient wood stove. *Courtesy University of Washington, Kinsey Collection.*

twice as big, that they would have twice as much speed and twice as much power. That was too much for Dave, who knew at least enough about physics to be aware that speed is gained at the cost of power and power can only be gained at the cost of speed. Dave roared, looking over the side of his bunk, for Slim to shut up. Slim was appalled. Here was a young punk who needed to be taught some respect. Slim warned that he would pull Dave out of his bunk and thresh him. Dave responded that he would be right down.

Dave was 21 years old, stocky and short-armed, but strong as a bull. Slim was about 26, tall and long-armed. Dave's feet hit the floor before he had finished saying he was coming. The two men went at it. Slim had several inches advantage in reach, and this was obviously not his first scrap. He began drawing blood at once, his long reach getting to Dave over and over while Dave tried to tag Slim in vain. But the worm turned. A wild charge brought Dave into close quarters with Slim, and he grabbed him. There was only a brief scuffle before Slim hit the floor on his back. Dave had him. He had Slim's arms pinned to his sides and held there by a knee-scissors. Then it was Dave's turn and he went to work. He gave Slim a pair of resounding slaps across the face, and asked if he would tell the men he was a jackass. The answer did not come speedily enough, and another energetic

Typical cookhouse, two flunkies, kitchen in background. *Courtesy University of Washington Library, Kinsey Collection.*

slap was administered. Slim mumbled acquiescence. Another brace of slaps, hard enough to loosen teeth, were administered for good measure.

Considerable blood and mucus exuded from Slim's nose. Dave got up, returned to his bunk and began to read. Slim, a crushed, humiliated figure of a man rolled up his bedroll, took his suitcase and disappeared into the night. He went to another bunkhouse. In the morning he changed tables in the cookhouse to get as far from Dave as he could. He evidently had no ideas of revenge, which in itself was quite typical of loggers of that period; he had been whipped, he had suffered a loss of face that was almost more than he could stand, but there was nothing to be gained by trying to prolong the affair. It was best forgotten. And Dave did not gloat over his victory. He never mentioned the subject in camp. Again, typical of the loggers and the hard, violent lives they lived, none of the other loggers said much about the incident either. It was over, a thing of the past, and there was the present and future to be thought about. And none of those men knew just how much future he might have left.

Sometimes matters took a more serious turn. There was Curly, for example. Curly was a vile talking loud-mouth. He was a bully by nature and, like most bullies, he skirted the edge of danger of physical encounter as closely as he dared, always trusting in his ability to run his bluff. It had become Curly's special sport to harrass one man in camp who could not

Second and seventh from left are uncles of the author.

reply in kind to his depraved insolence, because he spoke very little English. Franz was a Czechoslovakian. He had not been in the United States long enough to acquire much of a vocabularly in English, but he could work. His understanding of the language exceeded his ability to use it, so that when given simple, plain instructions he could usually carry them out. In the rigging he could *see* what was to be done, and he always held up his end of the work. He had been putting up with Curly's foul abuses for weeks without resistance. Afraid to take offense at something others might have considered harmless banter, and unable to know for sure, Franz did nothing.

It was Saturday morning. A bunch of the fellows were up early, shaving and dressing in clothes not quite so typical of their occupation. There would be a train into town right after breakfast. Looking forward to a day and night in town, where there would be moonshine and women, there was considerable speculative and lascivious talk during the dressing and shining of shoes.

Curly had nothing on but his socks. He pranced, naked, over to where Franz sat on the side of his bunk. Standing in close proximity to Franz, Curly loudly made a vile comment. There was no way Franz could be mistaken this time. He had been degraded, insulted, outraged. Franz still did nothing, but after the men had finished their dressing, while the clamor of the triangle rang throughout the camp, they all trouped to the cookhouse, all except Franz.

He didn't go to breakfast. He went to his suitcase; unwrapping the cloth around it, Franz inspected a wicked revolver. He opened a box of cartridges and loaded the gun carefully. Then he went to the door to wait.

As the men returned from breakfast, Franz stood quietly beside the door of the bunkhouse, one hand inside his shirt. It was noticed by some that he had a peculiarly grim look to his face. He just stood there motionless, without the slightest change in that look of doom in his eyes. Then Curly came. Franz announced quite calmly that it was time for Curly to die. Out came the hand from the shirt and Curly found himself looking into the muzzle of a gun that must have looked five times as big as it was. Franz began pulling the trigger methodically. He was not quite quick enough for the sudden movements of the flopping, leaping, bounding, cartwheeling Curly. From the second Curly saw the gun it would have required a machine gun to follow the movements of his hyperactive body with any chance of hitting the terrified, horrified gymnast once or twice. The first bullet had nipped a hole in the sleeve of Curly's shirt; after that the target was moving much to rapidly in too many directions. By the time Franz's gun was emptied, Curly was flashing down the railroad track with the speed of a gazelle. He certainly must have broken the existing sprint record of all time.

Franz turned to one of the men, handed him the gun, and asked him to call the sheriff. Told of the incident by telephone, the sheriff came as fast as he could by car and speeder. It took him four hours, at that, to get to camp. He wanted to know where the man was that was shot. Big Jim, the hooktender, answered that he wasn't sure if Curly was shot or not, as no one had been able to find him. The sheriff said they had better look some more, as he might be in the brush somewhere, dead. One of the men said Curly was lively enough when last seen.

Before the sheriff had left camp with his prisoner, the mystery was solved. The time came for the afternoon baking in the cookhouse. The cook began heating his ovens. He went to the flour bin, which refused to tip out from under the counter. The bin had been stuck before, so the cook gave a *hard* pull on the bin and it tilted out heavily and hit its stops with a thump. There was Curly, folded up like a jack knife. Even without the flour all over him, Curly would have been white from the terror that still gripped him.

Franz was not kept in jail very long. Several of the men had told the sheriff the whole story. Curly left the camp that day and never came back. I doubt if any other man ever went through a greater change than he did. I saw him fifteen years later and he would not think of saying a word to anyone that could possibly offend.

The cook in one of the big camps could make as much as the high climber. He was in absolute control of his domain, the cookhouse. At the 12-foot range, the cook, the second cook, and the kitchen helper could prepare breakfast for 125 men or more very quickly. In front of the range there was a long metal-covered serving table. As fast as the ham, the cooked cereal, the hotcakes and the eggs were done, they were put on platters or in serving dishes on the serving table by the second cook and the kitchen helper. From there, the flunkeys loaded their hands, and even their forearms with dishes and platters, and took off almost on the run into the dining room. The flunkeys were so adept at this work that it was unusual for one of them to drop or spill anything, even though a flunkey might carry as many as six platters at once.

As soon as there was enough food on the tables, the triangle was rung by the flunkey nearest the front door. The loggers would swarm into that door like a herd of hungry animals, and animals would have been as well mannered as some of them. The flunkeys would fly back and forth from the kitchen to the dining room. A platter would scarcely hit the table before it was empty. Those men would eat like there was no tomorrow. Many showed little regard for others. With two pork chops left on a platter, one man would take both, while the man sitting beside him had none.

Zig, the head flunkey in one of the Mason County Logging Company's camps, had a little more than his share of temper. Seeing the piggish table manners day after day had begun to wear on him. One day he set a platter of pork chops on a table, and watched while the whole platter of chops was slid off onto the plate of a burly bucker. Without looking back, the man passed the empty platter back over his shoulder to Zig, who took it and came down with it hard over the man's head, breaking the heavy platter and knocking the man out cold. The bucker slumped to the floor, and Zig went to the kitchen for more food. I think the only reason the bucker did not attempt to retaliate, after he came to, was that he was afraid he might have to whip some of the other men at his table too.

From his bakery, the cook always had plenty of fresh bread, cookies, pies and cakes. Coming back to camp Sunday night by speeder, it was pleasant to go to the cookhouse before turning in. No matter how late the men got back from their weekend of celebrat-

Part of a typical logging camp. *Courtesy William D. Jones, Historical Collection, Aberdeen, Washington.*

ing, they knew there would always be plenty of hot coffee, everything for sandwiches and cake in the kitchen which was never locked.

The old cookhouse looked pretty much the same, no matter which camp you were in. The building was made from rough lumber, battened against the wind and rain. There was no ceiling. The inner structure that held the building together, and the roof up, was in plain view. The walk leading to the cookhouse, the steps, the doorsill, and the floors themselves were all worn down from the champing of the caulk shoes.

The cook worked terrible hours. He got up at five in the morning and with few breaks worked until seven at night. The flunkeys didn't have it so bad. They had to get up early too, but after the breakfast work was over, the dining room cleaned, and the tables set up for lunch, the flunkeys had some time off. Again in the afternoon they had a couple of hours. Occassionally the train would be heading for the woods at about the right time so the kitchen help could ride out and watch the logging, while the engine pushed the empties through the bypass track and dropped in behind the string of loaded cars to prepare for the trip down. It was something to see the precision of the "tong pullers" (loaders) when they ran at a log, each with a pair of heavy steel tongs over his shoulder, while the Duplex puncher gave them slack, to throw the tongs deftly on the log, and jump clear as the donkey puncher lifted the log and let it down so skillfully that half the time he placed it exactly where the head loader wanted it on the load. The Duplex was an ingeniously engineered donkey; basically, the controls were just two levers. The puncher held one in each hand, and they operated the two lines that went through two blocks high above the center of the railroad and down to the tongs. Pushed forward, the levers lowered the tongs; pulled back, they raised the tongs. There have been a lot of loading methods used since logging began, but none that ever beat the Duplex in the hands of a skilled operator. Loaders liked it because, if the puncher was good, their work was cut in half. He could generally even shake the tongs loose after the log was in place, letting them swing back to the loaders who could go for another log without even climbing on the load.

The kitchen help liked those trips, to one side one day and to another the next time the train timing was right. The trip was not so much fun in the winter, unless there was only a couple of riders, so they could keep out of the way of the fireman, and the engineer would let them ride in the cab where it was warm and dry.

The trip *up* the mountain into the logging woods was fun, but the trip back was a real thrill. The geared locomotive usually had all the train it could handle on the steepest of the grade. Logging cars in those days had only hand-operated brakes, which the brakemen had to set by sticking an oak club through the brake wheel for leverage, and pulling the brakes as tight as they could. The kitchen crew, if they were in the cab, could see the concern on the face of the engineer. Some of those mountains were pretty steep.

Soon the train would pass through camp where the flunkeys would get off. Sometimes the train crew would come into the cookhouse and gulp quick cups of coffee, because that crew was another that worked long hours.

Once a week the whole cookhouse had to be scrubbed down. For that operation, all the kitchen help, excluding the chef and the second cook, turned to. Buckets of hot, soapy water were prepared and the mop brigade began to work. After the scrubbing, the floor had to be mopped dry. There were no squeegees then, and they wouldn't have worked on those battle-scarred floors anyway.

Cooks sometimes got tired of the long hours, and the seven-day weeks, and in spite of the high pay would dream of turning to other work. It was that way with Jake. During the brief periods when the camp had been shut down because of heavy snow, Jake had managed to get a large chicken house built on his place at Black Lake. Finally he was ready. Another shut-down came along, and Jake ordered a thousand baby chicks, having decided to quit cooking and go into the chicken business. Shortly the chicks arrived, and Jake put them in the chicken house. That evening he went out to see how the new arrivals were doing. The night was cool and, to his horror, he discovered that his chicks were piling up in the corners in danger of smothering the bottom chicks. Jake spread them out, and ran into the house to call a friend who knew about chickens. He was told that his chicken house needed to be heated and that screen should be placed a few inches above the floor over the chicks so they couldn't pile up. Those modifications would take time. Jake had to stay up *that* night stirring the chicks, and the next night. Already he was having second thoughts about the chicken business. He realized that he knew nothing of raising chickens. After he had talked to others who had tried, and failed, to make a living in the chicken business, Jake reluctantly found a new home for his thousand chicks and got out of the chicken business faster than he got in. Bordeaux's Camp Seven was going back to work, and back went Jake to his cooking.

CHAPTER 6
GRAYS HARBOR

When Captain Robert Gray discovered Grays Harbor, late in the 18th century, it could not have occurred to him that out of that harbor a century later would come thousands of windjammers, bound for many parts of the world with cargos of the finest lumber ever to be produced anywhere.

During the latter part of the 19th century and the early part of the 20th century, Pacific sailing ships plied the waters of the Pacific from San Diego to Puget Sound. Their cargoes were lumber, loaded in out-ports and dog-holes all up and down the coast as well as in the navigable harbors such as Coos Bay, the Columbia River, Willapa Harbor, Grays Harbor and the numerous ports in Puget Sound.

In 1851, Andrew Pope, William Talbot and Captain J.P. Keller had formed a partnership for the purpose of establishing a steam sawmill on Puget Sound. It was two years before the Pope and Talbot mill was ready to go, at Port Gamble, Washington. In fact, Captain Talbot took a load of lumber from Henry Yesler's mill in Seattle, which had started in March of 1853, to deliver to San Francisco on his ship, the *Julius Pringle*, before the machinery for the Pope and Talbot mill arrived on the *Foster*. The demand for lumber schooners increased. Hans Bendixson's shipyard on Humboldt Bay began operation in 1865, and built more than a hundred lumber ships in the 30 years that followed. Lumber ships in those days were mostly barkentines and schooners. Shipyards sprang up all up and down the coast. On Grays Harbor, John Lindstrom at Aberdeen and G.H. Hitchings and Matthews at Hoquiam, built wooden ships of both sail and steam.

With the coming of steam-powered ships had come other uses of steam. John Dolbeer, in California, had also developed the steam donkey to use for yarding logs. By the turn of the century donkeys were in use in nearly every part of the timber belt of the Pacific Northwest. Horses, and even oxen, continued to drag their logs over skidroads in many places, but it became increasingly apparent that where the donkey engines were used, log production was high. Many of the small logging operations continued to use horses or bulls, simple because they could not afford to buy a donkey engine.

Sailing ships were not quickly replaced by steamships. The construction of a sailing ship was such that it could not readily be converted into a steamship, so the windjammers continued to ply the Pacific until, one by one, they were lost in storms, or on rocks, or went aground while trying to make the entrance to a harbor. The steamship had the distinct advantage of shallow draft. It was also much more maneuverable and could go in and out of harbor entrances directly against the wind and without the assistance of a tugboat.

Between 1850 and 1900 mill after mill sprang up on both sides of Grays Harbor. It couldn't have been otherwise. Here was a good harbor surrounded on three sides by the finest timber in the world, in such quantities as to stagger the imagination. The timber stood tall; it did not stand in masses of acres or square miles, but covered multiple thousands of square miles. For the droughers, the barks, barkentines and schooners that habitually challenged the moody Pacific, loaded below decks and above until

Schooner Vigilant in Grays Harbor. *Courtesy William D. Jones, Aberdeen, Washington.*

The Janet Caruthers, aground shortly after leaving Grays Harbor, loaded to capacity with lumber. *Courtesy William D. Jones, Aberdeen, Washington.*

the hardy ships wallowed the ocean with the waterline feet below the surface of the water, and the skipper could hardly see to navigate his ship, Grays Harbor was closer to the big lumber market in California than was Puget Sound, and the quality of the lumber was as good as could be had in the world.

Often one company would own the timber land, operate a lumber mill, and own the ships that carried the lumber to the market, controlling the resource from the time it stood rooted in magnificent grandeur in a forest of the Olympic Peninsula, or on a timbered slope up the North River, to the time it was nailed in place as lumber in a dwelling house or industrial building far to the south.

Another company might specialize in one phase of the industry: one operating a mill but purchasing most of its timber; another logging the timber and rafting it to be towed to the mill that would pay the best price for the logs. Often the logging company sold logs to more than one mill: one might be a shingle mill where its operator wanted cedar; another might want only Douglas fir because most of his market was for that specific lumber.

One company that began to bore into the forest of

Schooner Edward R. West, loaded to capacity, above and below decks, leaving Grays Harbor. *Courtesy William D. Jones, Aberdeen, Washington.*

Typical Grays Harbor timber to go to the mill. A Lima Shay geared logging locomotive.

Polson's log dump and booming grounds; Hoquiam River. *Courtesy of William D. Jones Historical Collection, Aberdeen, Washington.*

the Olympic Peninsula long before the turn of the century, about 1885, was the Polson Logging Company. This progressive logging outfit was always in the forefront. Considering loggers as human beings with as many needs as other people, Polson's was the first to provide such things as metal beds with real mattresses instead of straw ticks, sanitary toilets with constantly running water, shower baths, and flowers.

At no time did Polson's have the reputation for highball logging that so many other outfits had, and deserved. Polson's produced 175,000,000 board feet of timber per year by 1917, but had the capability of logging 2,000,000 feet per day, had they chosen to pursue the hard-driving methods employed by companies that seemed to be motivated by only one purpose, getting logs, as many, as big, and as fast as possible. Sizable patches of timber were often left behind because they contained only timber that would make one or two rather small logs per tree. That would not have been so bad in itself, had any care been taken to preserve those smaller trees for future cutting, but when a few hundred of the big trees could be logged through a patch of smaller ones, knocking the smaller trees down as they went and leaving them to rot, the waste seems almost criminal. It was often the view of logging management that either taking the smaller trees, or taking the trouble to save them, would entail a financial loss.

Alex and Robert Polson had been in the logging business less than a quarter century, having started in a very small way, when the company was already one of the largest and most prosperous of any of the companies operating in the Grays Harbor area. By 1916 Polson's had rolling stock that included seven locomotives: two Shays, one Climax, three Baldwins, and a saddle-tank Porter. In addition, and in conjunction with the railroad equipment, the company owned two powerful tugs for moving their own rafts. These were the *Pilot* and the *Petrel.* There were two barges, one of which was a rail barge which transferred rolling stock from the mainline railroad at the harbor to Polson's own road up the Hoquiam River, where the company's booming-grounds were.

At that time Polson's had 55 donkey engines, mostly Puget Sound Iron Works engines. They were using six or seven North Bend overhead systems and had left ground-yarding to go entirely to high lead yarding.

Polson's was logging the country up the Hoquiam River, and across the mountain in the valley of the Humptulips River. The company had built its own steel bridge over the Humptulips downstream to connect with their railroad on the Hoquiam, and reach the booming ground. There was probably a total of about 75 miles of railroad, all well built roadbed and well engineered grades using 60-pound steel. There was no better logging railroad in the Pacific Northwest, and by then, 1916, there was no bigger or more complete logging operation in the Northwest either.

That is the way it was at the Polson Logging Company in the early part of the 20 century. The men of my family liked to work there because they had seen enough of straw ticks, bedbugs, lack of decent sanitary facilities and slave-driving. They were not alone. Loggers who had drifted from camp to camp all over the Northwest found Polson's so superior in so many ways to most of the other outfits that a strange phenomenon occurred. Many men stayed with Polson's year after year. The affectionate expression, "Papa Polson" developed. A logger who had spent a few days, or a few weeks, in town might be heard to say, "well, I guess it's time I was gettin' back to Papa Polson".

Strong feelings of seniority developed in those who had worked for Polson for many years. If an argument developed in the bunkhouse, about when the Big Creek country was logged, or where the best spruce had been found, the "old timers" always knew, or thought they did. One man might ask another how long he'd been around there. If the man said ten years, the first would sneer, saying he'd been there twenty. Then a third might join the argument saying neither of the others knew what they were talking about because he had been there before the road was pushed into Camp 5, which was more than twenty-two years earlier.

One of Polson's typical "roaders" in use at Camp 5.

Dumping loads into the Hoquiam River, Polson Logging Company. *Courtesy of William D. Jones Historical Collection, Aberdeen, Washington.*

Polson's locomotives in a line (12 of them). *Courtesy University of Washington Library, Kinsey Collection.*

Too big for tongs, this log had to be loaded with chain girdles. Polson Logging Company.

Pole road along which logs might be dragged a mile by a "Roader", using a "butt hook" to pull from behind. Polson's Camp 5.

Train of one-log-loads. Polson Logging Company.

Some nights in the bunkhouse tales would be told that were only lightly seasoned with fact. One of the popular subjects of many varied stories was the Olympic cougar. It happens that cougars grew to greater size on the Olympic Peninsula than anywhere else. They had been known to attack and kill children and, while I know of no case of one having killed a man, I do know of a big one dropping out of a tree onto the rump of a horse being ridden by a man, and when the horse sprang forward, running wildly, the cougar dropped off tearing a sheet of the horse's hide loose so that it draped over his tail. Certainly the cougar was an animal that was usually given a wide berth by everyone, and was often talked about.

One man told a graphic tale about having discovered a monster cougar in the tree above him when he had bedded down for the night. According to him, he carried only a shotgun, and while he wanted to kill the cougar, he knew he couldn't do it with birdshot. He said he whittled a spruce knot to fit in a shotgun shell, combining the powder charge from two shells and forcing the knot in on top; then at daylight he shot the cougar, and found that a splinter of the knot had gone through his heart. Silence reigned for a full minute while the men in the bunkhouse wrestled with that story, wanting to believe there might be some truth in it, but quietly doubting there was any. One man called Old Charlie settled the question for them by coming back with a story of his own.

With a far-away look in his eyes, he observed that when he first came he was walking along the trail when he heard a slight sound above him. Looking up, he saw nine cougars all lined up on a big limb. Old Charlie waited a minute or so before he added that it wasn't much of a night for cougars at that.

In the Polson camps almost anything from the sublime to the ridiculous to the remarkable might be encountered. Recently I ran onto an article in an industrial publication dated in 1933. It interested me, so I copied it. Here it is:

"The locomotive that pulled the first Northern Pacific train to the Pacific Coast in 1883 has been repurchased by the Northern Pacific Railroad for an exhibit at the World Exposition in Chicago. The en-

Yarding crew and one of the typical logs of Polson's Camp 5.

Loading scene, Polson logging company.

A Nash-Quad truck. Hanson used this to transport veneer blocks from his works to rail. *Sketch by Joseph H. Pierre, Jr., Monterey, Calif.*

gine, named 'Old Betsy', has been owned and used by the Polson Logging Company, of Hoquiam, Washington, for the last forty years."

It was on the north side of Grays Harbor that I first saw truck logging done with any success. North of Copalis Beach, about two and a half miles, there was a stand of big spruce. A good market developed for good quality spruce. A man named Hanson brought in a Nash-Quad truck. It drove with all four wheels and it steered with all four wheels. It was a brute for power, and it was a good thing for Hanson that it was.

From the stand of spruce to the open beach was about three-quarters of a mile. There was no road. It was then about a mile and three-quarters up the open beach to the "gap" where there was a steep planked approach to the county road at Copalis.

The spruce was cut into eight-foot "veneer blocks" that often had a diameter of more than six feet. They were loaded on the Nash-Quad crosswise two blocks per load and hauled to Copalis Crossing which was four-and-half miles beyond Copalis over gravelled county road.

There were two hazards to the operation. Even with the heaviest block loaded in front, the truck might rear up and lose its load on the steep plank approach. The other was the tides, which not only ran high on the beach at high tide, but also filled the river basin full where the truck had to approach the county road. I've seen that truck down in the river bottom, loaded to the limit with big spruce, in water up to the hubs with only wet, yielding sand under its wheels. If it hadn't been a four-wheel drive with plenty of power, it would have stuck. The thing was certainly maneuverable, steering from all four wheels, but loaded top-heavy as it was, the way it weaved and swayed up the road I always half expected it to tip over before it got to the mainline railroad at Copalis Crossing.

Another truck that hauled logs out of the Copalis area was a big, ugly, solid-tired, five-ton Bulldog Mack. It had a high cab and a sloping hood, and no

Some of many loading operations at the nearest rail to much valuable coast spruce.

Log so long one Bulldog Mack goes forward and the other backs up to haul it. *Sketch by Joseph H. Pierre, Jr., Monterey, Calif.*

Big log on a "Bulldog" Mack at Copalis Beach, Washington.
Sketch by Joseph H. Pierre, Jr., Monterey, Calif.

windshield. Because of the looks of it, it was called, even by its manufacturers, the "bulldog". It pulled a ten-ton trailer and hauled timber for the Northwestern Logging Company, which was operating near Copalis.

One day the yarder brought a spruce log to the landing that measured 14 feet in diameter. The loader put it on the Mack, where it was chocked and chained securely. The plank road from the landing to the county road creaked and groaned under the weight of the truck and its monstrous load. The truck driver thought that once he had reached the county road he would be in good shape. He was wrong. The huge load had not gone a quarter-mile before it broke through the road. Log jacks were used to raise the truck enough to block under it, and the truck continued. It broke through the road again, and again it was lifted and blocked. It was a costly process getting that log from the woods to Copalis Crossing. It went through the road nine times in a little over three miles.

The trouble was not yet over. Having at last gotten the log to the railroad, Northwestern was informed that the railroad could not handle a log anywhere near that big, it went at least four feet beyond existing clearances, both in width and in height. It was almost heartbreaking to see a beautiful "stick of timber" like that split, but it had to be cut into quarters before it could be shipped.

Early Packard trucks and trailers, hard tires, no cab. *Courtesy of University of Washington Library, Kinsey Collection.*

In my lifetime I've heard my share of "big log stories", and have no right to doubt any of them. I've heard of logs 16 and 17 feet through, but I have never seen a log bigger than the big spruce of Northwestern.

The "Bulldog Mack" also hauled massive loads of lumber from the mill at Copalis to the crossing. The mill was on the north bank of the Copalis River. A wooden-timbered bridge had to be crossed immediately after leaving the mill. Half the population of that little town would stop everything to watch the Bulldog Mack go by with its big load, only one day it didn't get far. I was near the bridge when the big truck started across, its chains and sprockets whining gutterally as it crawled up onto the bridge. Right in the middle of the span, there was a crack, a groan, a splintering, rending roar. The bridge, the truck, the load of lumber and two men dropped 25 feet into the Copalis River. The load of lumber began floating downstream. Both men, one with a broken leg, swam to the floating load and called for help. The old Bulldog Mack was out of sight on the bottom of the river.

A donkey engine was moved down river a mile from where it had been working. A diver went down and hooked onto the frame of the Mack. The donkey got her out on the bank all right, and another truck was brought in to tow her away for overhauling, but we at Copalis never saw her again.

Bordeaux headquarters and mill. *Courtesy Custom Photo Service, Littlerock, Washington.*

CHAPTER 7
CAMP SEVEN

The big mill at Bordeaux, headquarters of the Mason County Logging Company, was at the foot of the Black Hills in the State of Washington. The nearest town was Little Rock, six miles to the northeast. Beyond Little Rock, another ten or eleven miles, is Olympia, the State Capital

The logging operations in the Black Hills, shut down for awhile because of snow, were ready to start up again. In that part of the country it was uncommon for snow to come that heavy, as the Pacific Ocean is only fifty miles away.

Going to Bordeaux on New Year's Day, we went by way of Black Lake, up the Waddle Creek road past where the little town of Delphi once was before the logging moved so far into the hills that no reason was left to maintain a post office or store. Skirting the

Black Hills we missed Little Rock, crossed a little corner of prairie and headed straight west to Bordeaux. It was early in the morning, the air was crisp, and the green timber was fresh and beautiful. I never would have guessed, as we wound through the lower slopes of the Black Hills that morning, passing through magnificent old-growth Douglas fir, that one day I would be fighting a raging forest fire there.

Arriving at Bordeaux, I was surprised at the size of the mill, the town, and the shops where the company made up a lot of their logging equipment, repaired and even rebuilt donkey engines, and took care of the maintenance on the several locomotives they owned.

Standing on the tracks by the mill was a 110-ton Baldwin locomotive, big for a logging engine in those days. The camps were far back in the hills; one, Camp Five, was even a quarter of the way around the range of hills. Geared Shays, Heislers, and other engines suitable for the steep work, operated only between the landings, were the logs were loaded, and the camps. From there on, the big Baldwin made up trains and pulled them to the mill.

I hoped we would go out on the Baldwin, but we didn't. A Heisler, from Camp Seven, had been in the "roundhouse" during the shutdown and was now ready to go back. The fireman had steam up as we got aboard the "crummy" (closed car for men), the engineer arrived and we were soon moving off into the hills. The old Heisler was slow, but powerful for its weight. Behind the "crummy" was a string of 25 or 30 skeleton cars. The specially built cars were just two four-wheeled trucks joined by a heavy 36-foot timber, and fitted with heavy channel-iron log bunks and cheese blocks to keep the logs on the cars. The cars were joined together by "drawheads" similar to those still used.

A typical incline and four discontinued spar trees still standing: right center, center, and upper left. *Courtesy University of Washington Library, Kinsey Collection.*

Baldwin locomotive #6, Mason County Logging Company. The company road was once called Black Hills and Northwestern Railway Company. *Courtesy Pete Replinger, Shelton, Washington.*

Up the grade we went with the old Heisler making a lot of fuss, but not going very fast. We wound around in those hills for about 25 minutes before we finally reached the summit. From there we dropped into a high valley several miles long. Eventually we came to a switch, where the track went straight ahead to Camp Four, and went up another narrow valley to the left to Camp Seven. The train came to a halt, the brakeman opened the switch and we began to move up the narrow valley.

Interested in the passing scenery, and particularly in the high ridge of virgin timber on our left, I didn't realize for a moment that the man beside me was talking to me. He asked if I had ever been to Camp 7 before. I told him it was my first time in that part of the hills. He assured me I would see something unusual. I asked what, curious at his enthusiasm. He said he was talking about the incline. My curiosity growing, I asked what an incline was. He only said I would soon see. Loggers were like that. A man didn't bother to introduce himself, if he felt like talking he just spoke his mind.

On went the Heisler, creating a tumult in the little valley that echoed back from the hills. Around the last curve, we pulled the cars off onto a siding. The brakeman set the brakes, we got out of the "crummy" and crawled onto the Heisler's tender. The locomotive pulled out of the siding, leaving the "crummy' with the log cars. A few hundred feet ahead, the incline came into view.

It looked like a railroad track standing on end, that incline. Nearly half a mile above us, on the big ridge, was a donkey such as I had never heard of. It was called a snubber. At the bottom there was a curve into Mason County Logging Company's mainline; but extending straight ahead of the incline, tracks continued across the tiny valley to the side of the opposite hill. The switch was always kept open to that "runaway track" except while log loads were being switched into the siding off the mainline. The degree of incline was close to 45 degrees. That meant that to go up the track, for every 100 feet the locomotive went forward it had to go *up* 45 feet. I couldn't understand how they expected to get the engine up the mountain, yet that was where it had to go, because Camp Seven was up there, and so were several billion feet of valuable old growth timber.

I asked someone who had crawled up onto the Heisler's tender beside me what would happen next. The logger answered that the block car was coming down. I looked up that ladder-of-a-railroad and, sure enough, a small structure on wheels was creeping down to us. The lower end of the incline and the upper end were connected by telephone, and one of the brakemen had called for the block car. The engineer pulled the Heisler onto the relatively level track that joined the end of the incline and in a little while the block car was down. Ignoring the drawheads, the brakemen secured the engine to the block car by two fat cable straps made of the biggest cable obtainable.

The snubber, powerful as it was, could not pull the Heisler up the incline by itself. The fireman had reduced the fire, but had a head of steam. Too hot a fire could cause the boiler to blow up on the incline because most of the water would be at the back of the boiler.

Hooked up and ready, the snubber engineer was told to start the pull. The old Heisler reared up on end, and with the combined power of the snubber on top, and the locomotive below, we started to climb. Quite a few of the men were better satisfied to go up the incline on foot than to ride, with that much weight dangling from the line of the snubber. With our feet braced against the back of the skirt of the tender, we could feel the surging power. Looking up the incline I could see the heavy cable with its bight around the block car, so taut the line was in the air straight as a string from the block car to where the line disappeared over its rollers at the top. It took only a few minutes to reach the top. During that time I learned why the engineers had built the "runaway" track at the bottom.

Incline Slim once was taking three loads of logs down the incline to add to the string below, waiting for the road engine. A drawhead broke. The weight of the three cars of logs suddenly dropping on the safety strap, broke the massive cable from which it was made. The three cars were loose. The sudden lurch caused Incline Slim to fall off the back end of the back car, where he was riding. Scrambling to his feet, Slim

A two-truck Heisler, the same model as the one owned by Mason County Logging Company.

The donkey used for snubber, Mason County Logging Company. *Courtesy Custom Photo Service, Littlerock, Washington.*

saw the tension of the snubber's line snap the block car 100 feet back up the hill, and turn it over on its back. Looking down, Slim saw the fastest moving railroad cars he had ever seen. One crooked log on top was lifted off by the wind and tumbled off down the canyon. In the middle of the hill, the tracks moderated in grade for about 200 feet, then continued downhill at the same grade as above. At the point where the reduced grade again increased, Slim could see daylight between the wheels and the rails. It was at that point that the crooked log blew off from the sudden renewed drop in addition to the wind. So straight was the track that the wheels of both trucks of all three cars took the rails again perfectly after they crossed the hump. Nobody knows how fast those cars were going before they plowed into the opposite hill, but they did just that. No attempt was made to salvage the cars. There was nothing left to salvage. Looking down there, as we heaved along up the incline, I could see ends of splintered logs still sticking from the hillside. Then we were at the top, and I was looking at the most monstrous machine I'd ever seen, the snubber. It was awhile later that I met Incline Slim, and we became friends.

My first job at Camp Seven was as a flunky in the cookhouse. Eventually I worked at a number of different jobs: digging donkey settings, pulling whistle, bucking for a big Willamette two-speed, splitting for the same donkey, and learning the hard way how much wood a donkey like that can consume, and finally rigging trees for new landings. But it was while I was still in the cookhouse during off hours that I got acquainted with Incline Slim, who as his name implies was in charge of the operation of the incline. I also met the engineer of the snubber, and he explained the workings of the monster to me. One of the largest donkeys ever built, and *the* largest in the Pacific Northwest at that time, the mainline drum was nearly eight feet in diameter, the lower portion of it dropping below the donkey's deck. Water was piped to the specially constructed heavy-duty brakes for cooling while loads were being snubbed down the incline. The upright boiler must have been six feet through, and twelve feet tall. The smokestack extended up another seven feet. Built by Skagit, the thing was a giant masterpiece, yet no bigger than necessary for its job.

The donkey initially got up the hill by its own power, the crew securing its mainline to big stumps and pulling until another stump farther up was needed. After it was in position at the top, all the rest of the machinery for the entire camp was pulled up by it, beginning with the ties and rails as the incline was built.

Here again the money saved from grading out and laying the miles of railroad that would have been required to reach that high country in the usual way, made the snubber pay for itself many times over before the Camp Seven timber was logged out.

On top, there were miles of railroad going in several directions, over which the Heisler went from one landing to another, bringing the logs from each side to the snubber. Back into the high country southwest of the camp, the railroad was crooked and steep. It was decided by the engineer and the camp superintendent that it would be best for the Heisler to *follow* the loaded cars down instead of leading them. It was a wise decision, considering the poor braking systems of that time. So a passing track was built at each landing, to allow the engine to pass the loaded cars, drop off the empties which had been *pushed* up the hill, and come in behind the string of loads.

One day the Heisler started down the mountain, behind a string of 17 loads of logs. The brakes had been pulled so tight that, at the landing where the grade was slight, the engine had to strain to start the train down the hill. In a little while the grade steepened, and the train began rolling along without the push of the engine. By the time the train had rounded two or three curves the brakes had warmed

Two Lima Shays. Mason County Logging Company. *Courtesy Capital Museum, Olympia, Washington.*

Climax locomotive, Mason County Logging Company. One of many and varied geared engines used by Mason County Logging Company. *Courtesy of Custom Photo Service, Littlerock, Washington.*

up, and the grade was steeper yet. Smokey, the engineer, began to add the braking power of the engine to keep the train under control.

The train was heavier than usual. It continued to pick up speed. Smokey began to fear that he would not be able to get the speed down to a safe level again. Down the track, half a mile ahead, was a sharp curve to the right followed by a switchback to the left a little further on. Smokey did the last thing he could do, he pulled the Johnson bar to the reverse position, and cracked the throttle. There was no way in the world a locomotive would brake better than using that technique. Smokey waited, worried, and sweated. The train squealed, bucked, and swayed as it rumbled down the track, still picking up speed. Smokey knew at last that the train was a runaway. There was no way it could stay on the rails through what was ahead. Smokey waited no longer. He began a series of short blasts on the whistle. It was the signal for the crew to leave the train; 17 cars ahead, the brakemen left the train end-over-end. The fireman left the cab and picked himself up out of the dirt. Then Smokey left; he tried to light on his feet with his legs flying, but the speed of the train was far too great and he also did several end-overs before he came to a stop.

Smokey climbed up on a low bank beside the track, sat down and put his head in his hands. Up the track the fireman was approaching, limping. Down the track, the brakeman was coming back up. Smokey could still hear the train. Then he heard a prolonged, prodigious roar. He knew it had happened.

The fireman climbed up on the bank, sat down beside Smokey, and pulled up his pant-leg to look at an injured knee. Neither of them spoke a word. The fireman would have liked to say something to ease his engineer's anguish, but he asked himself, what do you say to a man who has just lost the locomotive that was the pride of his life, to say nothing of the probable damage to 17 log cars. The brakeman had stopped on his walk up the hill, and cocked his head as if listening, then he continued to approach the two on the bank. The brakeman stood on the ties in front of Smokey. He offered to go up to the landing to tell the boss but Smokey said he must already know, he must have heard him whistle the runaway.

Little more was said for three or four minutes. All three were disturbed about the wreck. Smokey was devastated. Twenty-two years he had run logging engines and in all that time he had never damaged an engine. He had lost fewer logs in transit than any lokey engineer in that part of the country, because he

The old "Six-Spot", wrecked April 1913, one man killed, Bordeaux, Washington. *Courtesy Custom Photo Service, Littlerock, Washington.*

was always careful on crooked, uneven tracks. Now he had lost a whole train. None of the three gave any thought to their personal good fortune in having escaped without serious injuries. Their only thoughts were of the train they knew was piled up somewhere down the track.

A faint whisper of sound was coming from somewhere down the track. Could it be escaping steam from the wrecked locomotive? Had someone in camp heard the whistles and the roar of the wreck? Had someone started out with a speeder to investigate? The sound, now slightly louder, was not the sound of a speeder.

Suddenly Smokey sat up straight. There was something familiar about that sound. Then, to the supreme amazement of the three men, the corner of the old Heisler poked around the bend a quarter of a mile down the track. The fireman looked at Smokey in disbelief. Smokey jumped off the bank like it was hot. He ran to meet the lazily sauntering locomotive, his heart in his mouth. It couldn't be, but it was. Behind the locomotive was one carload of logs. When the wreck had occurred, the derailment of each successive car had tended to slow the rest. The second car ahead of the locomotive had been the last one to leave the track. As its front end dropped from the grade it raised its rear drawhead right out of its counterpart on the following car. The locomotive and one car of logs were still safely on the track, detached from the wrecked train. The engine, as Smokey had left it, was in reverse with the throttle cracked. Lazily it began to puff its way softly back up the hill. All Smokey had to do was swing aboard, the Heisler wasn't travelling over two miles an hour.

Later a mobile crane was brought up to salvage cars and logs. Damaged cars were loaded on undamaged ones to be hauled in for repair. The logs were loaded again. So ended one of the most bizzare accidents in the history of logging. Smokey had experienced the proof that truth is stranger than fiction. The company had not always been so lucky. Years before the Big 6 Spot had left the rails, and turned upside down, killing one man.

The Camp Seven story needs *some* introduction to the most important man in camp, Walt Monroe. Walt's father, Matt Monroe, was the superintendent of

Mason County Timber Company's Camp Four. Walt and Matt had one expression in common, no matter what the age of any man in their employ, they always addressed him as "lad" or "my lad". In stormy, bitter cold weather Matt would turn to the man nearest him and observe wryly, "What I mean to say, my lad, she's kinda bad". A cocky cousin of mine with an inflated opinion of his prowess in the logging woods once approached Matt for a job in the rigging. He asked Matt if he happened to need a damn good man. Matt said he did, and asked the cocky one if he happened to know where he could find one.

With Matt, Walt had been around logging, except for the time he spent in school, all of his life. He never wanted to be anything but a logger. He got no special treatment from his dad. He worked his own way up; it was inevitable, and his chief ambition was that he should work his way to the top in the woods.

Thomas Jefferson once had much to say about the "natural aristocracy". He meant that one did not have to belong to the royalty, the nobility, or the very rich to be a person of true quality. I have known a few men in my life who were members of that natural aristocracy. They were the ones who made things happen. Let an emergency of any kind occur and watch the people. It will be fortunate if one of those present has some of the qualities of the natural aristocracy. He will know what to do. Better yet, he will assert the needed leadership to get it done. All those present will instinctively bend to the will of such a man.

Walt Monroe was just such a man. He was a natural leader and a skillful and intelligent logging operator. None of the men, up to and including the "high rigger", questioned his judgment. Of course, nobody is always right, but Walt Monroe came as

Landing, donkey, car-loads of logs. X-mark above Matt Monroe. Mason County Logging Company. *Courtesy Capitol Museum, Olympia, Washington.*

Enclosed speeder used to transport men from camp to headquarters, but also used to rapidly get injured men to the nearest ambulance. *Courtesy Custom Photo Service, Littlerock, Washington.*

close to that state of perfection as anyone I ever knew. He could be gentle, or hard as steel, as the occasion required. He never did spend much of his time around the camp office, leaving the work to the riggers and the hookers. He saw to it himself that the logging went forward as it should. He was right out there, whichever side happened to be in the toughest "show" or operating on the most dangeous ground would be where you would find Walt. When things got better there, he would go to another side to see how things were going. His presence was always welcomed. He had a way of straightening out problems, ignoring the trivial, and going hard after that which was important.

Walt was phlegmatic. He took everything as a matter of course. Nothing ever made him lose control of his emotions.

Walt was there the day one of the loggers got his tin pants caught on the point of a pair of tongs. When the donkey puncher got excited and lifted the man off the ground, shook him loose and let him drop six feet into the thick mud, there could have been a killing. The loader got up and looked around for a club to use on the donkey puncher. The first thing he grabbed turned out to be the tip of a tree that was buried in the mud. Failing to break the tree, the loader ran wildly back and forth until he found something small enough to wield, and big enough to be effective. Lifting the big, gnarly club of a limb high over his head, the loader ran for the donkey bent on mayhem. But the donkey puncher had seen the light. He did not stand by while the loader busied himself finding a weapon. He left that donkey far behind before he stopped, ready to run some more if necessary. After all, one just doesn't stand still while a crazed loader approaches on the run with a large club. As I recall, Walt said "all right, lad, put it down." The loader seemed to come to himself. He stopped, continued to swear with abandon, but he put the club down. When the puncher thought it was reasonably safe, he returned to the donkey, ready to apologize to the loader for the unintelligent caper. He was not surprised when the "old man", as Walt was respectfully and affectionately called (he was still in his thirties), gave him a concise, one sentence lecture on the hazards inherent in becoming rattled while in control of a machine of such power.

Walt was there the day a man was injured in the woods, and brought to the landing more dead than alive. While the others did their level best to do all they could, it was Walt who remained completely calm, no matter how he may have been feeling inside, until the man was safely on his way to the hospital.

Walt was there the day a certain young logger got too much freezing weather and snow. The young buck had no tin pants or weatherproof coat to keep out the wind. He did not have a pair of good Churchill rigging gloves. He had a second-hand pair of "corks" with broken soles. He seemed to be freezing to death. Finally he just said, to hell with it, he was going to camp. He was the youngest member of the rig-up crew. Walt gave him a steady look, but didn't say a word.

The young logger warmed up a little during the long walk down to camp. He went to the office to let the timekeeper know he had quit. He was told that Walt had to sign his slip, and that he would be down at noon. Noon came, and the young fellow saw Walt go into the office as he came out of the cookhouse. As soon as Walt had gone back to the woods, the boy went back to the office. He asked if the old man had signed the time slip. The timekeeper answered, no. Angry, the boy demanded to know what he meant, and if he had even *asked* him to? He was exasperated beyond measure. Doggedly the timekeeper insisted that he certainly had asked Walt to sign the slip, and he just didn't do it. I guess I may as well admit that the sorry excuse for a logger was myself. I announced that I wanted some clothes fit to work in, because I was going back to work in the morning.

The next morning I was with the crew in the "crummy" on the way out to the woods. Walt was already up there, some problem at the side had brought him out earlier by speeder. I showed up with the rest of the rig-up crew, expecting to be ordered off the job the minute Walt saw me. He was giving some instructions to the boss of the rig-up gang when he turned and saw me. He gave me the same steady, searching look as he had the day before when I flew off the handle and announced that I was leaving. He didn't say a word. That was the difference between Walt Monroe and the average woods boss. Walt knew I didn't really want to quit. He knew I was young, and he intended to go more than half way to see that I had the chance to make it there, if I had the guts to stay. He had deliberately refrained from signing my time slip, in hope that I would reconsider. Here is the proof that he was really a big man. I wasn't some valuable man that he just couldn't get along without,

Innocent of snow, these hills seem benign compared to the same railroad in the dead of winter. *Courtesy Pete Replinger, Shelton, Washington.*

This cedar required a 14-foot saw. A close-up of what one of these trees looked like. *Courtesy University of Washington Library, Kinsey Collection.*

Long logs loaded on two cars with one between; logs of unusual length were sometimes shipped on special orders. Mason County Logging Company. *Courtesy Capitol Museum, Olympia, Washington.*

he could have managed without me very well, but he was thinking of me, where the average woods boss would have dismissed the whole thing quickly, and said let him go, we're better off without him anyway.

Walt was there the day two of the men got into a foolish argument at lunchtime. Men were sitting all around on stumps, on the donkey sled, and on logs. One of the two got in such a rage because he couldn't make his point, that he threw his hat down on the ground, jumped off the stump on top of it, stomped it ten or fifteen times with both feet, and then gave it a wild kick. The hat sailed over to the roots of the stump where Walt was sitting. Deliberately, and calmly, Walt laid down his sandwich, got down off his stump, stomped on the hat some more, and kicked it back to its owner. He climbed back up on his stump, picked up his sandwich and began eating. He didn't say a word; he didn't have to. The object lesson had done its work. It is doubtful that any man present ever made such a ludicrous ass of himself again, certainly not around the "old man".

The many men who worked for Walt Monroe had one thing in common, they were loggers. Aside from that, they were very much individuals. One man was a double-jointed acrobat who liked to show off his agility around the bunkhouse in the evenings. He was a capable logger. He went to the hospital once, and the receptionist needed certain information for the

Industrial Accident Commission. She asked for the names of his children. He couldn't remember them all. She asked how many children he had. John hesitated before answering, but finally settled on the number 12. It seemed that John didn't get home very often, and about every second time he went home he found a new baby. He had sort of lost track.

One man was called Whittlin' Bill. He could make anything with his pocketknife: chains, bells, figurines, and almost anything he happened to think of. He was the fireman on one of the yarders.

Another fellow, whose first name was Martin, always liked to be in town when the carnivals came through. He liked to take on the carnival wrestler. He usually won. He was a choker-setter.

Then there was Incline Slim, a man of skill and responsibility in camp, a man's man. I met him one night unexpectedly at a Black Lake party. The party was at a cabin on the lake, far removed from any neighbors, which was an excellent thing. The party was well supplied with stolen turkey, home brew, and moonshine. Slim showed up dressed too well for a logger on a weekend bust. He was a tall, nice looking guy, in a strong, masculine way, and his voice was half a dozen tones lower than any I had ever heard. Slim had had a couple of belts of moonshine before he took notice of a girl he thought required attention. He introduced himself, and shortly had an arm around her waist while they discussed possibilities. The man, Les, who had brought the girl, tried to put an end to the negotiations, first by verbal remonstrance, then by a wild haymaker aimed at Slim's head. He should not have done that. Soon people were wondering where Les had gone. Carl said something very much like "I saw him runnin' down the road, right after Slim hit 'im". Sandy asked Slim why he hit him. Slim said offhandedly, and in a voice like the rumble of a lion, that he only hit him a light tap because he kept pestering him, adding that he didn't want to hurt him. Those who witnessed the brief, unequal encounter, unequal not by size and weight but because a "city feller" had tried to take on a hardened logger, swore Slim's "light tap" had knocked the man ten feet, flat on his back.

All these men, rough, tough, powerful, and sometimes vicious, yes and sometimes kind and sometimes gentle, Walt Monroe handled with such firm, masterful skill that he kept order in his camp better and more consistently than most teachers can in a grade school. Trouble was bound to crop up once in awhile, but it didn't last long if Walt was around.

I didn't leave Camp Seven until word of my sister's death. A dozen miles or so don't sound impressive if you are going to ride a train or speeder, but when my brother and I got the news of her death, in Centralia, there was no train or speeder available. Instead of walking the track, which would have been the long way, we started hiking across the hills. Neither of us had ever walked out straight across country before. After some hours of making our way through standing timber, and crossing hill after hill of logged-off land, we began to hear donkey whistles. We were puzzled because we should not have been approaching any active logging show.

Coming at last to the top of a ridge, we could see the donkeys working, and we could see a logging camp. Aware now that at some point we had gone down the wrong side of a ridge, it nevertheless seemed best to continue to the camp, not only to get our bearings, but to possibly find transportation to where we could catch a train, or "stage" (bus) to Olympia.

It was rough going down that last ridge. We knew there must be an old railroad grade there somewhere, but we didn't hit it. We bulled straight down the hill, jumping from one old down-snag to another, and floundering through brambles and brush. We headed for the camp instead of the logging side. Eventually we reached it, and crossing a railroad, we went straight to the office, knowing there would be someone there. Clarence asked the timekeeper what outfit it was. The man answered that it was Mason County's Camp Five, generally called Gibson Creek. He looked at us, curious that we hadn't known where we were. Clarence explained that we had crossed the hills from Camp Seven, and that our destination was Bordeaux. He told us we were in luck because the road engine would be there to take a train out in a few minutes, going to Bordeaux. We walked down the track to where we could see a string of loads on a siding.

The Big Baldwin soon arrived; it was the 9-spot, mate to the 6. We climbed up on the tender and rested from our rough hike while the engineer coaxed the train around through those hills and over the summit. Then it was downgrade all the way into Bordeaux. Though we hadn't planned it that way, it worked out all right, and we hadn't lost any time.

It was while we were hiking that day that I first began to realize what tremendous inroads had already been made into the timber of the Black Hills. There had been times when most of what we could see were logged-off hills. I came face-to-face with the realization, that day, that the time would come when there would be no more such timber as we were used to seeing. Since then, the Black Hills have all been

Logged off hills such as these were not designed for easy hiking. Much of the Northwest now looks much like this. *Courtesy Joe Bell, Tillamook, Oregon.*

logged off. That huge, seemingly inexhaustable supply of beautiful, big old growth timber has been all logged out.

Not long ago I ran across that old friend, the 110-ton Baldwin rod engine, in a collection of old locomotives. It had been restored and painted. It looked just like it did when we used to ride her to and from the camps. Feelings of nostalgia and regret swept over me. In my mind then, as today, the old Baldwin is at the head of a train of logs, thrusting with its powerful pistons as the drive rods swept briskly, propelling the drivers along the rails while the deep roar of the stack and the squeal of wheels on rails echoed back from the hills as we pulled up the crooked track toward the summit of the Black Hills.

I can still picture the sight and sound of the barking, grinding, roaring two-speed swing donkey as the puncher poured the power to the mainline coming up the hill dragging a ten-thousand board-foot turn of logs, in the air most of the time, over bumps, and hollows, and stumps. I can still see the puncher of a yarder straining his eyes toward his taut line as he endeavored to keep the turn moving though it whaled into stumps along the way, sometimes so hard as to knock the stump half out of the ground. I can still see the little cluster of choker-setters fighting chokers onto logs to make up the next turn for the donkey. I can still see the men scramble for safety as the rigger shouted "Ho!" and the whistle punk, with his whistle wire over his shoulder, threw one quick wave into the wire, and the donkey answered back "Woo", the rigging began to move, coming suddenly tight and jerking the logs half off the ground as they began their trip to the landing.

I fought nearly every phase of that battle as a young man. Things have changed now, but I sometimes wish I could go back in time and once again take part in the most sweaty, sunburning, freezing, miserable, glorious work in the world.

Typical forest fire. *Courtesy Joe Bell, Tillamook, Oregon.*

CHAPTER 8
THE FOREST FIRE

The next summer, most of the logging camps were shut down because of extemely hot weather and low humidity that constituted grave fire danger.

The condition prevailed for five weeks through late July and the entire month of August. Many of the men from camps around the Eastern side of the Black Hills were in Olympia, biding their time until logging opened up again. Many of the men were already out of money, and were picking up such odd jobs as they could find. Day by day we expected a change in the weather that would allow logging to begin again ...day after day there was no change.

Then one day late in August, the sky darkened. It was not the sign of coming rain that we saw in the sky. Everyone knew what it was, a forest fire had broken out.

Word soon came that the fire was along the edge of the Black Hills bordered by the valley of the Black River. The fire warden needed a lot of men. He needed them fast. The danger of the fire penetrating the hundreds of square miles of virgin timber was very real. Trucks came to town to pick up volunteers. With several hundred unemployed loggers in town, it was no problem for Charlie Roundtree, the fire warden, to find all the men he needed.

Trucks loaded with men, axes, saws, shovels, and fire-trail hoes convoyed out of town to the west to Black Lake, on to Delphi, and out the Waddle Creek road into the timber. At a point directly up the hill from the main fire, the trucks unloaded the men and their tools. Ray Bordeaux had agreed to provide food for the fire fighters for as long as necessary, and well he should. Much of the endangered timber belonged to his company. Many of us had come out in our own cars. My cousin, Pat, had driven his car and I had come with him. We were all told to leave keys in the cars in case the fire should get into the heavily timbered area through which the road wound, and the cars had to be moved in a hurry. The warden had said it, and at that moment his word was law. He even had the power to stop cars and order the drivers onto the fire line if he had to.

We scattered out along the line of fire. While some scalped the dry moss and duff off the surface to bare the ground with hoes in a long winding strip (called a fire trail) between the fire and the green timber, others began falling snags, many of which were already burning, to prevent them from burning off and falling the wrong way, throwing burning chunks as they fell to further expand the fire.

All of us were out there in clothes we didn't care about and wearing our caulk shoes for safety as we would be jumping from one burning log to another and without the caulks might slip and fall into the fire. As we left the road, up in the timber I saw Ray Bordeaux pull a new Buick off the road and head for the fire. That was just his *brush* car, Pat commented, adding that he had a *good* one that he used on the highway. I was surprised how neatly dressed Ray was, and wondered if he was really going to fight fire, and what he would look like by night if he did.

Jumping from one burning log to another, running through smoldering limbs, moss and bark in between, we reached and fell (the proper term is "felled", but I don't want some old logger to kill me for using it) as many of the snags as we could, as fast as we

Fire crowning out. *Courtesy Joe Bell, Tillamook, Oregon.*

with the fire had burned holes in our clothes. The soles of our caulk shoes dried and shriveled from the heat. Caulks loosened and dropped out. The shoes were almost unbearably hot on our feet at times, and many feet had blisters, though we tried our best to stay out of the fire. By the end of the second day, the situation began to look hopeful. Time after time we had contained the fire and were patrolling the fire trail with men covering each hundred yards for three miles. Just as we would think we had it whipped, the fire would flare up and the rising wind would carry burning materials across the trail. While some of us were trying to get one "jump" under control, the fire would jump somewhere else. It wasn't until the third day that everyone, including the warden, was convinced that we had the fire under control.

We were all wrong. The east wind increased, and fires that had almost died flared up to blaze fiercely again. The cry that the fire had gotten away was relayed up the fire line, it was moving rapidly up the hill. The fire had run into a stand of saplings which ran from the point of the jump all the way to the tremendous old stand of old growth fir. The flames were soaring fiercely in the pitchy limbs of the sap-

Destruction left by forest fire. *Courtesy Joe Bell, Tillamook, Oregon.*

could. We were working along the edge of an old logged off area, a natural fire hazard with dead logs, limbs, bark, leaves and needles forming a tinder-dry ground cover.

There was no let up to the work other than to take gulps of water whenever a water boy happened to find us. sometimes we couldn't wait for the water boy, if our work happened to take us to where a trickle of water came down the hillside. We just dropped to our faces in the sooty stuff and drank it.

After twelve hours on the line, each man was allowed to catch a little rest, up the hill away from the fire, but the pay went on from the time we got to the fire until such time as we should have it under control. I don't believe any man rested more than three or four hours. He did his eating and resting at the same time, lying on the ground for a few winks, bone tired. There was no chance to wash except to splash a little water on our faces when the water boy came around. In two days we all looked like men of another race, and exceedingly dirty ones at that. Unavoidable contact

Millions of acres of fine timber like this have been destroyed by fires in the past. These trees will average 8 feet through and 250 feet tall. *Courtesy Oregon State Forestry Department, Salem, Oregon.*

Donkeys in the path of the Tillamook fire were banked with earth to prevent sleds from burning up. *Courtesy John Coats, Portland, Oregon.*

Destruction in its wake, fire still rages. Here, the sun was not seen for 8 days. *Courtesy John Coats, Portland, Oregon.*

Fire-killed timber must be salvaged within a few months before sap-rot and worm damage render it unprofitable to log and market. *Courtesy John Coats, Portland, Oregon.*

lings. It was just a matter of time before it reached the big timber. Cries ranged back and forth up and down the fire trail. Someone yelled to move the cars. The cry was passed from man to man down the trail. It didn't matter who got to what car first, keys were in all of them. Pat and I were at a point nearer the timbered road than some of the others, so we raced to reach the road.

Near exhaustion from running up the trail, we came to the old Gray just in time to hear the most terrifying roar ever heard. We knew the fire had crowned. The roar was almost deafening. Heat from the crown fire penetrated the forest clear to the road instantly. Jumping into cars we started motors to join the caravan of fleeing vehicles that poured northward through the timber. I was one of the last to start.

We raced north as fast as we could go while the fire raced west. It came up the hill like a roaring tornado, killing everything in its path instantly. The crown fire led the way at fifty miles an hour, and the ground fire fed by burning, falling branches, followed somewhat more slowly.

Before I got clear, the crown fire passed directly over the car, 200 feet above. Roaring and flaming hundreds of feet in the air, it created such a violent up-draft that air was sucked under the bottom of the fire in a strong wind, bowing the saplings and vine maple almost to the ground. The heat boring through the top of the car was so intense that I feared the headliner would burst into flames. I wouldn't try to guess how hot it got in that car, I know it was all I could stand.

Two Mack trucks traversing desolate skeleton of what was beautiful green forest. *Courtesy John Coats, Portland, Oregon.*

Clear of the fire, I did not immediately look at the car to see the scorched paint. I was too intent on watching that sky-licking inferno race up the mountain faster and faster until, because of its very heat, it was jumping hundreds of feet ahead as it went. It didn't slow down until it reached the top of the mountain, when its up-draft, pulling air into the fire from the west side of the mountain, also created a check force that slowed the fire long enough for us to get in front of it and start over.

On the west side of the mountain (one of the Black Hills), the agonizing work began; falling trees uphill toward the fire, cutting the fire trail, putting out ground fires and, where feasible, backfiring. Everything known about fighting forest fires, at that time, was employed. Then nature came to our rescue, we got rain. Between our efforts and the effects of the rain, we finally did have the fire contained and under control in two more days.

I have fought fire since then, and have witnessed the most terrible one of all, the great Tillamook fire, but that fire in the Black Hills remains the most fearful experience I've ever had with fire. I don't need any more.

Tired, burned, blackened, and half in shock, we returned to our homes after five days intimate contact with the fire devil to draw our pay, which was considerable, and contact our respective logging companies to find out when they would start logging again. We knew they would. The fire hazard was gone, but so were 20,000 acres of the finest timber on earth.

CHAPTER 9
GIBSON CREEK

After a layoff or a shutdown, loggers in those days did not always return to the same camp they had worked at before. Gibson Creek, Mason County Timber Company's Camp Five, had a good reputation. I had liked the looks of the operation the only time I had been there with my brother months before. There were not so many miles of railroad to be traversed getting in and out of camp as most places, and it would be a change. So many loggers drifted from camp to camp in those days that one superintendent had once been heard to comment that he had three crews all the time, one coming, one going, and one working.

A few days after the forest fire, Pat and I headed for Gibson Creek. Taking the gravel road out of Oakville, we drove down to the parking area where the road ended at Mason County's railroad and left the car there. You could do that in those days, and come back to find your car unmolested.

We hiked the three or four miles up the track to Camp Five. At the office we found the Superintendent. His name was Joe Lameaux. Since *his* boss's name was Tom Bordeaux, I decided I'd better play it French, so when I introduced myself I pronounced my name Pea-air instead of Pier as it had always been pronounced. (The name Pierre was arbitarily shortened to Pier by an ancestor, and most of us went back to the authorized spelling.) The superintendent asked, are you a Frenchman? and I knew I was hired. We *both* were. We had a good supper and got beds. The next morning we were all over the old Shay and some were sitting on flatcars, their legs dangling over the sides, as the hardy old locomotive wound its way around the mountain to the high ridge where one of the logging sides was working.

We went to work there just as one yarder was ready to move to a new tail tree. This involved removing the skyline from the tail tree, pulling all the rigging from that tree, and moving the whole assembly to a new tail tree several hundred yards around from where the area had been logged out.

Even the spar tree would be dropped and cut into logs. One of the things that had to be moved was the water line to the yarder. Pat and I had been put to digging up the water pipes from where they had been buried under the skyline road, which was not a road but merely the path over which logs traveled, pulled by a big Skagit swing donkey. Whenever the logs happened to touch the ground, which wasn't often, there was the danger of breaking the water line so it had been buried more than four feet deep at the point where it passed under the skyline "road".

Pat had started to dig at one end of the pipe where it was exposed a few yards away from the road, and I had started digging on the opposite side, where the pipe came out of the ground. The source of the water, and the setting of the yarder, was such that there had been no other place to put the water line. We had been digging an hour, and had reached the point where we lacked only about three feet of coming together, when somebody goofed. The skyline, a steel cable two-and-a quarter inches thick, was cut loose from its anchor stump beyond the tail tree and came whistling down out of the sky to strike the ground just between the two of us with such force as to bury itself in the ground with a sickening, heavy *whop.* Any man who has worked around the woods has had his close calls, and if they are unavoidable he thinks of it as being all in a day's work, when that work is logging; but when the close call is the result of

Mason County Logging Company, Gibson Creek. *Courtesy of Pete Replinger, Shelton, Washington.*

someone's blunder one doesn't appreciate it much. That incident our first day there was our spontaneous introduction to Gibson Creek.

It was at Gibson Creek that it was again impressed upon me the wisdom in observing basic safety precautions. How many times I had heard the instruction "Stay out of the bight of the line". I knew every other logger had been cautioned the same way, over and over again. The haulback line, unwisely, had not been strung in a straight line from the yarder to the corner block. A line under stress that runs against a stump will act like a dull saw, eventually it will cut its way through. We had finished rigging the new yarder tree, and had logged a few days. One of the men was standing on a stump waiting for the rigging to stop before he proceeded to cross the path of the pull. The haulback had been sawing the back of a small stump, and finally cut its way through. The line whipped straight *hard* from the force of the yarder's pull. The man on the stump was turned the other way, unaware of his danger. The snapping line whipped over the stump where the logger stood, grazing its top so closely that it neatly cut the heels off both his shoes. It did more than that, it turned the man exactly upside down. He stood on his head for a second, then toppled off his stump, unhurt, but not likely to be found in the bight of the line again.

It was at Gibson Creek that I saw a Duplex puncher make one of those mistakes that every donkey puncher hopes never will happen to him. There were several logs piled at the landing ready to be loaded. The head loader wanted one certain log out of the pile for a load "peaker" (top log). He threw a pair of tongs on one end of the log and signalled the puncher to pull it from the pile.

The donkey puncher got "rambunctious" with his controls. The log was 40 feet long, and didn't come out of the pile easily. The puncher threw full power to the pull. The log came out and dangled from the tongs. The puncher had to set the log down for the other pair of tongs to be set, so he moved the tong-line lever to the "down" position. The end of the log came into solid contact with the ground, but as he allowed slack so the log could tip, the tongs simply came off the log. For several seconds we watched the tongs dangling helplessly in the air while the long log stood on end. Then, completely free of any control, the big log began to lean slowly toward the donkey. Faster and faster it came, as the donkey puncher jumped for his life, and everybody near scattered. The log came crashing down across the roof of the Duplex, and smashed it to splinters. Having wrecked the donkey structure, the log glanced off the boiler, fortunately not rupturing it, and hit the unfinished load. The car shuddered from the impact, and the logs already loaded came off the car in various direc-

10-spot, called "slobbertop". Both Lima Shays. Though it is numbered (1) the engine on the left is a later model lokey. *Courtesy Custom Photo Service, Littlerock, Washington.*

Black Hills and Northwestern Railway Company. #9 spot at Gibson Creek (Mason County Logging Company) *Courtesy of Pete Replinger, Shelton, Washington.*

tions. It was a good thing everyone near had run for safety, because any number of men from one to a dozen could have been injured or killed.

I guess that was the "bad luck landing". It was the same place where the skyline donkey broke the spar tree in two, at the buckle guys, dropping the twenty-ton spar tree, blocks and lines, straight down onto the landing, half of it ending up across the skyline donkey. Yet, no one was killed that time either.

It was at Gibson Creek that I got a badly needed new perspective of myself. All the loggers I had ever known had hated to see the college men come into camp. Often these men were in the woods for the first time, and by loggers standards they were inept, clumsy, soft and ignorant. There was one such man in the same bunkhouse with me. God help me, in my own ignorance I looked down on him. I thought of him as the "college punk", the same as the other loggers did. It never occurred to me that this man, working to pay for his next term of education, looked upon me as that vulgar, profane, ignorant and extremely cocky young man.

On one particular night it happened to be rather quiet in the bunkhouse. Several were dozing in their bunks, there was no card game going on. I eased over to where William, the college man, was quiet with a book. Without asking to be excused for the intrusion, I asked what he was reading. He answered that it wouldn't interest me. I saw the name, *Chaucer* on the cover as he closed the book. His certainty that I couldn't possibly appreciate his reading material stung me. Proud of the fact that I also could read, yes and spell too, I held my emotions under control as I said that I often read lots of stuff, when I got a chance, and had some good stuff at home for that matter.

It was a rather tense few moments, and I don't know how it happened, but in a week or two William would occasionally offer me something to read, bringing certain passages to my attention. What was far more important to me, I learned to respect William for what he was, and ceased to condemn him out-of-hand for what he was not. It was curious, both of us had been guilty of bias and a weird sort of chauvinism.

As our friendship developed, it was not uncommon for William and me to discuss portions of Shakespeare while some of the men played cards, some talked logging, others amused themselves with dirty stories. I enjoyed the discussions. Nevertheless, in the woods I was my old cursing, blasphemous self. It was a couple of years later that something happened to change all that, but that is another story for another time. But I quit thinking of college boys as "college punks" and I have William to thank for that.

It was at Gibson Creek that I got credit for courage, and intrepidity that I did not possess. Late summer had withered into fall, and fall had chilled into winter. We were rigging another new spar tree. Louis McDade, the high rigger, had limbed and topped the tree, and pulled the first small block and line up after him by a light rope. The "passblock" as it was called was secured to the tree by a light cable strap. From then on everything went up the tree, or was let down from it, by the pass line, including Louis McDade himself.

Louis was one of the finest high riggers in the country. When he was rigging a tree he was the absolute authority. Not the side rod, not even the superintendent, but Louis. He was not only the high rigger, he was the head rigger.

Operating the passline drum on the donkey, it was my job to pull Louis up the tree, hold him until he secured himself with belt and spurs, then he would send the line down for a heavier strap, to hang for the haulback block. One by one all the eight guys were pulled up and secured to the tree. A strap for the bullblock had to be hung and eventually the bullblock itself would go up and all the lines would be run through their respective blocks and it would be time to hand-drag the strawline into the brush to begin the woods rigging. Louis had no part in that. Riggin' slingers and the hook-tender, and their crews, took care of all that trivia.

It was a big undertaking to rig a spar tree. It was cold weather for hanging up there hour after hour. I kept my eyes on Louis constantly for a hand signal. with other sounds going on around, his vocal signals might not be understood. "Down" Louis yelled at lunch time, at the same time making the unmistakable hand signal. I eased him down the tree to the ground while he kept clear of the tree and kept himself straight by giving the tree an occasional kick with his feet. Louis uncoupled himself from the line and straddled towards me, feet wide apart because of the spurs. He straddled right up to me, and looked me straight in the eye and said something like "damn it, man, it's cold up there; when I say down I *mean down*". He unbuckled his spurs and shed his belt. We all went into the cookhouse for lunch because of the cold.

Climber topping spar tree. *Courtesy University of Washington Library, Kinsey Collection.*

All through lunch hour I thought about the incident. Louis was a tough logger, I knew that, but he hadn't cussed me out. I was glad for that. I had seen many men humiliated in front of the other men, by tough hooktenders who prided themselves on their evil

vocabulary. I felt self-conscious, even though Louis hadn't cussed me, because all the others had heard what he said.

Back at the tree that afternoon Louis belted and spurred himself up, hooked onto the pass line and gave me a thumb-up. I hoisted him up the tree smartly, and he threw his steel-core rope around the tree, fastened it to his belt and let the line back down. As he called for various items to be pulled up to him, one or two of the rig-up crew secured them to the pass line, and I lifted them up until Louis's hand shot out in the horizontal position, meaning "hold it right there". The afternoon wore away; Louis was nearly done with the rigging of the tree, but it was too late to finish it that day.

From 225 feet up in the air I heard him yell "down!" faintly, as he extended his arm and swept it downward. It was the moment, I suddenly realized, that I had been dreading. In my mind I could still hear the brittle ring of his voice when he said so vehemently "...I *mean down!*". I cut the brake almost loose and let the drum roll. Louis was coming down the tree, and *I mean down*. Watching him kicking the tree away from him I realized he was only kicking the tree about once every forty feet. He was dropping fast. I pressed the brake a little more, but too lightly. The man was dropping too fast, and I had to stop him because he was already perilously close to the ground. I came down on the brake harder, and Louis stopped. His stop, one foot from the ground, was so violent that his arms and legs flopped crazily. Trembling, I let him down the last foot and he got out of his gear. I think I heard some idiot mutter "I guess by God the kid showed ol' Louis!"

A typical old yarding donkey.

Straps were often used instead of tongs on big logs. Tongs might tear out and damage the timber. Also would be dangerous. Loading here done under a gin pole rather than a spar tree. Crotch line with spreader bar. *Courtesy William D. Jones Historical Collection, Aberdeen, Washington.*

Bucker halfway through log. *Courtesy William D. Jones, Aberdeen, Washington.*

I had no intention of showing "ol' Louis" anything. In fact, if I had any intentions at that moment they concerned running, I was sure I was in for it. I stood at the drum tensely, watching Louis. When he began to go about his business, without making a move towards me, the truth dawned on me. He, and all the men present, believed I had given him that fast flight down from the tree purposely, to show him just how fast I *could* let him down, in retaliation for his loud complaint that morning.

The last I saw of Louis McDade was when he got married and brought his bride to camp where he had secured one of the family houses for them to live in. Everybody was determined that Louis and his wife would get a charivari that could be heard in Oakville ten miles away. We got together all the pans, tubs, and old saws we could find to beat on and went down the track to Louis's house late that night. Somehow, Louis had given us the slip. The joke was on us, but we didn't know it until we had charivaried an empty, lighted house for thirty minutes, then forced our way in to find... nobody. Louis never did get his charivari. He took some days off and the two of them went somewhere. Then more snow with high winds came, and the camp closed for awhile. After so much time had elapsed, there didn't seem to be much point in a charivari any more... if indeed there ever was.

Seven men sitting in undercut, no crowding. Note fallers springboards in place. Mason County Logging Company. Courtesy Capitol Museum, Olympia, Washington.

Whitney's "Big Jack" (Hammond Tillamook). *Courtesy Lila Boge, Tillamook, Oregon.*

The "Molly-O", one of Whitney's well known locomotives; a Climax. *Courtesy Lila Boge, Tillamook, Oregon.*

CHAPTER 10
HAMMOND TILLAMOOK AND B.W.

When I went to Tillamook in 1927, I thought I had already seen logging on a big scale. My brother-in-law, Joe Carnahan, was already working there firing a locomotive, a big 230-ton Baldwin, the biggest I had ever seen used on a logging road. On the side of the engine were the words, *Whitney Company* and *Big Jack.* All their engines were named. A Climax was named *Molly O*, another engine was named *Andy*, and another *Min.*

I came to the operation while negotiations were under way for the purchase of all Whitney Company's holdings in Tillamook County by the Hammond Tillamook Lumber Company.

Other logging outfits that I had known had fine railroads, many donkeys, and many locomotives. The difference between them and Hammond Tillamook (formerly Whitney) was that this company not only had all those things, plus a huge mill at Garibaldi, but it had its own steamship line.

The formerly huge timber holdings up the Kilchis River were about logged out. The camps were soon vacated and much of the machinery was put in storage. Headquarters Camp at Idaville was not closed immediately, as all the steel remained to be pulled from many miles of railroad grade. But the new company began getting most of their timber from the upper Miami River and in the mountains beyond.

During the readjustment, another man and I built a dike on the Miami River, which took a few months. Then I went to work for the Coats Lumber Company which had its own rail logging camps south of town. Coats at that time had two mills, one at Yellow Fir and one in Tillamook. This company had their own lumber schoooner, the A.F. Coats. In the mountains along the coast, they had high, steep hills to negotiate. There were several high trestles and one unusual one called the "crib bridge".

In later years, even after forest fires had run through the area, damaging many of the trestles, truck roads were built of planks using the old railroad grades and some of the trestles to bring out remaining timber.

Before I had worked for Coats in the railroad logging very long, I had the chance to go with Hammond Tillamook, longshoring at Garibaldi. For the first time I was with a company that did its own logging, its own milling, and its own shipping. Their most profitable and important operation at that time was the Hammond Line steamships.

The mill at Garibaldi was large and lacked nothing that went to make a complete mill operation. resaws, dry kilns, planing mill, dry storage, and acres of dock space, all were there. I know that several of the pioneer logging and lumber companies of the Pacific Northwest had their own lumber fleet, and up the Columbia the innovative Simon Benson had even built sea-going "cigar rafts", up to 1000 feet long and containing 6,000,000 board feet of timber. But Hammond Tillamook was the first company I had worked for that did its own logging, railroading, milling and shipping on such a scale.

The Tillamook Harbor bar was so treacherous, in spite of the jetty, that when the ocean was rough, we longshoremen used to stand by while the skipper of the *Tillamook*, or the *San Pedro*, or one of the others lay off the mouth of the harbor waiting for high tide to see whether or not he wanted to take a chance on running in. In such cases, when the captain finally made his move, whether on the first high tide or a later one, we would sit out there at Barview watching until the schooner squared away and made a run for the bar. Then it was the skipper fighting the breakers for his ship, heaving, twisting, pitching, corkscrewing, and taking water over the stern, the ship would

Hammond Tillamook mill, and ships, at Garibaldi, Oregon. *Courtesy Pioneer Museum, Tillamook, Oregon.*

Two donkeys loaded on cars behind locomotive ready to move to new side. Coats Logging. *Courtesy Lila Boge, Tillamook, Oregon.*

Schooner A. F. Coats in dock at Lahaina, Territory of Hawaii, 1914. *Courtesy John Coats, Portland, Oregon.*

plunge through the narrow channel only yards away from the rugged rocks of the jetty, until minutes later she was safe inside.

Almost without exception the longshoremen had been loggers up the Kilchis. All knew what a day's work was, but when the schooner finally tied up at the dock every man of us knew he was in for a real workout. The men worked in teams. Each team went to its regularly assigned part of the ship and lumber began to be swung aboard. Since the longshoreman's union was not yet in control in that out-of-the-way port, there was usually no stopping until the ship was loaded. We would work until ready to drop. It was important to Hammond's that the ship spend as little time as possible tied up in dock. We sometimes worked 30 hours without stopping, except to grab a couple bites off a sandwich or take a drink of water.

Joe and I had one of the tough spots to load, the port wing. That was our regular spot, and that's where the timbers went. The smallest we got were six by eight, the biggest twelve by twenty-four.

A sling load of timbers would swing over the hatch and come down. We had to pick up the timbers, a man at each end, and stack them according to length so that our stowage followed the contour of the hull and would stay in place when the ship pitched and rolled at sea. Joe had always been a powerful man and I was young and unusually strong for my weight. I guess that's why we were assigned to the loading of timbers. After about so many hours of lifting those timbers around, the flesh of our fingers used to begin to tear loose at the upper joint. Then we would have to let the edges of the timbers rest on the palms of our hands more than on the fingers. We knew they would heal before the next ship came in. With the overtime we could make more money loading one ship in thirty hours than we could make at anything else to be had around there in ten days or two weeks.

The Schooner A. F. Coats lying in the harbor of Hilo, Hawaii, with 1 million board-feet of lumber. *Courtesy John Coats, Portland, Oregon.*

Finally the longshoremen's union began to get control of all the ports. The companies, and many of the non-union men, fought back. Other unions became involved, there was a strike, and a shutdown. I left the Tillamook Bay area.

Booming around for a few months, I found myself in Portland without a job. I had been cutting timber and clearing land for Portland's Bull Run reservoir, but that job was done. By then I had a wife to support.

I had heard that Joe Carnahan had gone to work for the company that had bought the big Baldwin that he used to fire for Whitney, and later for Hammond Tillamook. I only knew he was somewhere in the coast range.

I went to one of the old time employment offices that used to flourish in the big cities. This one was somewhere near the west end of the Burnside Bridge in Portland. I had bought a second-hand pair of caulk shoes. My old ones were shot. The manager asked what work I wanted. Looking at the listings, I answered I supposed rigging. He spoke of Ostrander over in Washington and asked if I was familiar with that country, he indicated a listing for choker setters. I said I had never worked there. He declared it was good enough, a choke setter's paradise. I never knew there was a paradise for choker setters. Then he said there was the B.W. who would take another man or two. I asked where that was, and was told in the Coast range, near Jewell. I asked if it was on the Kerry line. He said Jewell wasn't on it, but B.W.'s line ran into it. I think I paid two-and-a-half for the job. I left my wife in Portland and took off hitch-hiking. I didn't have money to spare for "stage" fare. There was none that ran all the way into B.W. anyway.

By night I was in the B.W. camp. I had secured a bed in one of the bunkhouses and was on my way to the cookhouse to get something to eat, when I saw a big Baldwin that had a familiar look. I thought it was "Big Jack", but it was not the same engine. On the tender it read B-W Timber Co. On the cab was *65497* and on the headlight, 65. The engine was just returning to camp from the Kerry Line, and sitting in the cab was my brother-in-law, Joe, and now he was on the engineer's side.

Joe asked me to come to his bunkhouse after supper. The other members of the train crew stayed in the same bunkhouse because they got up earlier than the loggers. After greetings, Joe asked how I happened to come up there. I said I had bought a job in Portland, and that was where it was. Joe asked what I would be doing. I told him setting chokers. He wished it was something else. When I asked why, he said it was too dangerous and made old men out of young ones too fast. I asked what about loading timbers into the port wing? Wasn't that about as hard as work could get? He said yes, but with times so hard we had to take what we could get. I told him times were worse than they were before, that there were only two outfits listed, and B.W. was one of them. If they weren't so far from town, neither of them would have had an opening.

We talked about old times and family, but Joe didn't try to discourage me from fighting chokers any more. He was proud and thankful to be back railroading, especially to be running the engine. He told me what a great engine it was. After a few minutes I went to my bunkhouse.

At eight o'clock in the morning I was in the brush. Some changes had to be made in the rigging before we could start logging. There had been a dance in an

Two Climax engines crossing crib bridge with a big donkey, Tillamook County. Coats Logging Company. *Courtesy of Joe Bell, Tillamook, Oregon.*

The big Baldwin owned by B. W. Joe Carnahan, who had fired "Big Jack" on the Kilchis tracks, now appears at the engineers window. This engine was called "old 65". *Courtesy of Mrs. J. R. Carnahan.*

old schoolhouse not far from camp the night before. Where there was a dance, and where there were loggers, there was always moonshine. Some of the crew were not in the best condition to be in the woods that morning.

While we finished moving the tail block several of the boys expressed some dissatisfaction with the condition of their heads. The rigging slinger showed no mercy. Why should he? The choker setters were getting four seventy-five a day, the highest anyone was paying at that time. The rigger expected a day's work from everybody, no matter who or how many had made asses of themselves the night before with the country girls and moonshine.

We were finally ready, and the buttrigging with its chokers being in good position, we hooked onto three nice logs. We hurried to get clear. "Ho!" yelled the rigger. We waited. The donkey did not return the single whistle blast. "Ho!" sang out the rigger again, loud enough to be heard half a mile. Nothing. The rigger wondered profanely if the whistle punk was asleep. The rigger started crashing through the brush to where the whistle punk should have been in plain sight, but was not. The screams of the rigger could be heard a mile as he reached the place where he could view the peacefully incumbent form of the man, who had lolled back into the bunch of brush behind the log he had been sitting on. The rigger roared at the man, who floundered to his feet. The angry rigger ordered him to either pull whistle, or go to camp.

So began the day at B.W., a camp that I soon learned had one purpose, to get as much timber logged and loaded and put into the Columbia River as was humanly possible, with this reservation that B.W. was concerned with the safety of the men. If you could boil company policy down to a sentence it would be: "Get logs, if you want to keep your job. Do it safely if you want to keep your life."

We were in good timber and being a side just beginning, we were close to the railroad track. In fact, the yarder was so close to the track that the tree, rigged for loading too, had the two donkeys set within a few yards of each other. We sent those logs to the landing so fast that the loader couldn't keep up. The loaders had to catch up when we stopped occasionally to move the tail block so we could reach a new swath of timber.

Our side was on a railroad spur, so the road engine did not pass close to us, but I did have the occasion to watch the big engine take a train over "the hump" more than once. After a train of 47 to 50 cars had been made up by the geared engines, and stood ready at camp to go to the Columbia River, Joe would come in with a string of empties, drop them on the siding and back into the string of loads. With the brakes released he would back up until all the slack was out of the couplings. Pulling on the throttle, he would start forward with just the right amount of power so the engine would not quite spin its wheels and, taking the slack out of the drawheads as he went, the cars would start one at a time. Of course, the whole train was moving in less than six seconds, but the technique was important. If the slack had not all been telescoped before the pull began, the engine could not start the train. While the train was in the process of picking up momentum, the engineer had to be careful not to apply too much power or his drivers would spin, the train would slow and he would have to pick up on it again, adjusting the throttle to get the most possible pull without spinning.

Once moving along, though, the engineer could really pour the power to the drivers and they wouldn't slip. Often the train would be all the big engine could take over the mountains, one or two more cars would have been too much. Log loads are heavy; fifty cars of logs could weigh as much as seventy-five cars of general freight. These were the things a good engineer learned, and learned well. He learned what his engine could do, and exactly how to make her do it.

Logging went on. A lot of B-W timber went to the river. Finally there came a shutdown. Back in Portland, out of a job, I decided to go to Salem, where we had a few acres of land. We had a notion that somehow I could get a house built and have a place of our own to live in while the depression lasted. By then we had a daughter.

I scrounged around and got hold of enough lumber and bricks to build a house. I built it myself, chimney and all. We had to carry our water up the hill from a spring. We needed a well, so I erected a tripod, hung a pulley and rope, and began to dig. A little way down I hit hardpan and had to go to using half a stick of powder every little while, using a stumping auger to make the hole for each shot, to loosen the stuff. Working alone, I had to climb the rope hand over hand ot get out of the well every time I fused a shot. But the well got deeper and deeper, and I had to climb farther and farther until the day came when the depth of the well was too great, or I was too tired, or both. I had drilled down and tamped in three-quarters of a stick of dynamite with cap and fuse. I lit the fuse and started climbing. Near the top I began running out of steam. Only the certain knowledge that I had to climb out or be blown out enabled me to pull myself up that last couple of feet. I drew my legs up and threw them over the rim of the well, and tumbled out, just as the shot went off. It rained chunks of hardpan around there for a few seconds, but that last shot had done it. The well began to fill with water.

It was getting into spring, in the middle of the depression. I had traded work for supplies, a cow, an old car and other things we needed, including meat, potatoes, and other staples. One thing I did not have was money. I needed a job. There was none to be had. Every train that came through Salem was loaded with unemployed men. Every empty boxcar became a rolling refuge for them. Cars of lumber were populated in the ends and on top by men who were not hobos, though they stopped in the same hobo "jungles". They were not bums, though they were reduced to living by the same means, begging. They were just discouraged, beaten, displaced persons who were ragged, dirty and hungry, and who would have wept if *someone somewhere* had only offered them the chance to earn a bare living.

I was better off. At least I had a house and my family (we had two children by then) and had some staple foods, but I did need work and spent some sleepless nights trying to figure out where I could find it.

Big gnarly western cedar. *Courtesy Pioneer Museum, Tillamook, Oregon.*

Author's brothers-in-law, Ed and Clarence Yocom, at Independence works, 1910. Note springboards used to elevate "fallers" above the swelled butt.

CHAPTER 11
PINE COUNTRY

A friend of mine named Mike, whom I had worked with on a plywood show, came chugging up the hill to my house, south of Salem. He asked, still sitting on a homemade seat on the chassis of a Model T Ford, if I wanted to go to work. Incredulously I asked what kind of work, and where. He said he had heard from his brother in Medford that Lamm Lumber Company was going to start logging again after the first. He thought he'd go down there. I asked him if that was the place he had told me he used to fall timber, and he said that was right, but the only chance to get on was to *be* there when they started up. It sounded great to me so I asked him where the Lamm outfit was. He stated it was northeast of Klamath Falls, back in the hills fifty miles or more from the mill, which was at Modoc Point on Klamath Lake. I looked at his stripped-down Ford and asked if he intended to drive that there. He answered sure, that there was a road right into camp, and he urged me to come along. I asked about money, it would cost quite a bit for us to get there. He said not much, the car was light and got almost thirty miles to the gallon on the road. He added that he had about ten dollars. I said I had some chittum bark about dry that I could take down to the Jew, which ought to help.

We went into the house and drank coffee, and talked some more. It was Wednesday, we could be ready to go by Friday. Mike would pick me up at my house Friday morning. That would give me a chance to sell my bark and get things in shape for my family, until I could send some money home. We were ready to leave Friday morning, when Mike said I should not forget my guitar. It didn't make much sense to me at the time, but I got the guitar and put it in the little load we had wrapped in blankets and tied down, because the skinny little platform behind the seat with no back, had no sides either.

The roads in those days, even the main ones, were uphill, downhill, and crooked. That little Ford ate up the miles like I wouldn't have believed. By Saturday night we were at Lamm's camp. We put our gear in a bunkhouse, just like we owned the place. Sunday we went back down the road for a swim in the Williamson River. It was sunny, pleasant and refreshing. Monday morning we were at the door of the office before it opened. Men were milling around, anxious looks on their faces. I was sure I had no chance of getting a job. There must have been twelve men there for every job. I *wouldn't* have gotton one either if it hadn't been for Mike.

When we finally got a chance to see the superintendent, he recognized Mike. He asked if Mike was there because he wanted to fall some more timber, and Mike answered that that was what he was hoping. The boss told him to draw his tools from the tool house, that the crummy would be going out in a few minutes. He started to turn to someone else, but Mike wasn't through with him yet. He asked Claude, the superintendent, if he couldn't find something for his partner, meaning me. The big man gave me a quick once-over. Then he looked me in the eye and asked what I could do. Lying nimbly I said I could do everything, but that right then I'd do anything he had for me to do. Claude Houghton, the superintendent, thought for two seconds, then said the track needed a lot of work and I was to go out on the track with the section foreman and help him a few days until he could find a place for me in the woods.

So, for the next three weeks I learned some things about one of the most surprising stretches of railroad in the country. Beginning near Kirk, Lamm's railroad went straight across Klamath Marsh and nearly 20 miles up into the mountains covered with sugar pine, white pine, ponderosa and other kinds of pine timber.

When the road was built, the engineers thought by going directly across it would save the cost of constructing railroad all the way around Klamath

A mountain in pine country. This was the area of Lamm's operation for many years. Williamson River in foreground.

Before #3 left the factory at Eddystone, for delivery to W. E. Lamm (Baldwin 2-8-2). *Courtesy Klamath County Museum, Klamath Falls, Oregon.*

Men, center bottom, give perspective to these ponderosa trees.
Courtesy Favell Museum, Klamath Falls, Oregon.

A ponderosa pine. W. E. Lamm, right. *Courtesy Favell Museum, Klamath Falls, Oregon.*

Marsh. After the construction crew had dumped millions of cubic yards of rock and dirt into the virtually bottomless quagmire, they began to have second thoughts, but by then they were committed to the project, and were too far into it by several million dollars to take another course of action. The end result was a railroad grade, across the marsh, that continued to settle year after year, and into which went a fortune in materials and labor each year for maintenance.

The other wild thing about that stretch of railroad was that it crawled downhill. Because of the altitude, temperatures at night went to freezing or below all summer long. In the daytime it would run from 80° to 100°. As the rails shrunk during the cold of the night, they contracted downhill. The full length of the railroad, from the marsh to the landings was on a grade, so the rails were constantly on the move. Rails were shortened with a cutting torch near the bottom, and after awhile there was a supply of rails of various lengths which could be interchanged as needed. The crawling process was aggravated by the passage over it of long heavy log trains. On one occasion the train conductor was sitting up in the cupola of the caboose when the track under the caboose kicked sideways so violently that it threw him down out of the cupola. Picking himself up, thinking the caboose had derailed, he looked back in surprise to see a big kink in the track. It had to be repaired before the train could return. Had the track kicked out while the engine was there, there surely *would* have been a derailment that would have been costly and dangerous to equipment and men.

I never thought I would ever be a gandy dancer. Loggers have as strong an aversion to track work as they do to mill work, and no self-respecting logger would think of doing either unless he was absolutely forced to. But, strangely enough, I found I didn't mind it. Maybe that was because I knew it was temporary. Anyway, I was paired off with a good-natured

W.E. Lamm's Baldwin, Mikado-type, with log train, #3 Lokey. Klamath County. *Courtesy University of Washington, Seattle, Washington.*

Irishman who really knew track work. We replaced ties, reinforced bridges and elevated sections of the track wherever settling had occurred. I learned to tamp ties, and drive spikes, with the best of them. The railroad work only lasted three weeks, for me, and I have never done it since. But it hardened my muscles and when the "old man" told me he had a spot for me in the woods, I was toughened up and anxious to get on with it.

My first job in the woods was hooking behind the turn-out cat. The whole process of logging, there, was new to me. First came the fallers and buckers, of course, and then came the turn-out cat and its hooker. The turn-out hook was a large curved hook which I would throw over the end of a log. The little Caterpillar tractor which was called the "kitten" would take one short pull. That would swing an end of the log out so it was free from the rest of the tree. Then came the choker-setters who wrestled chokers under the logs and "buttoned them up". Then came the big "cats" (and we really thought they *were* big then), the Caterpillar-60's, to drag the logs to the landing. The choker-setters coupled the chokers to the buttrigging and the 60's drummed the logs up under their track arches, or the "Paul Bunyan" wheel loggers, as the case might be, (we had both types there and the big wheels were eight to ten feet high), and the load of one to four logs, depending on size, was pulled to the landing, front ends up and back ends dragging.

At the landing was, to me, a freak of nature called a jammer. A jammer was a donkey engine that had legs. It arrived at each new landing on its railroad wheels, then it was jacked up on its big legs and its wheels tucked up so that log cars could pass right under it. A string of cars would be pushed through the jammer by the locomotive and let down car by car during the day as they were needed to load. The loaders did not pull tongs. The pine was loaded by *end hooks.* The two lines to the hooks came together above so that when each loader pulled his hook to his end of the log, by a rope attached to the back of the hook, the hook would bite into the end of the log. An end hook was constructed so the bit-end had a sharp cup-shape with a sharp cone-shaped prong in the center of the cup. The prong would bite into the log and the sharp rim of the cup would also bite in, just enough to prevent the prong from tearing the soft pine.

With five cats logging to the landing, things really moved. The loaders jumped around in the heat and dust like they were wild, from eight to five, except for lunch hour when they collapsed under the nearest

Back-cut finished, the ponderosa drops. *Courtesy Favell Museum, Klamath Falls, Oregon.*

Cat picking up load with arch. *Courtesy Bob Lewis, Klamath Falls, Oregon.*

Cat, Paul-Bunyan wheels, and pine logs. *Courtesy Favell Museum, Klamath Falls, Oregon.*

Cat pulling load up to landing under "big wheels". *Courtesy Bob Lewis, Klamath Falls, Oregon.*

Horse-drawn "big wheels" bringing logs to landing on right, car loading center, jammer left. *Courtesy Favell Collection, Klamath Falls, Oregon.*

pine tree to regain enough strength to finish the day. Load after load of fine pine lined up, ready for the Mikado Baldwin to take to the mill.

For some strange reason, even though the jammer was on the railroad and could have been converted to burn oil, many burned wood. There was lots of wood unsuitable for lumber, but good enough for fuel and it burned hot. Burning wood did not represent the same loss of dollars that might have been the case in fir country. Between the pitchy smoke from the jammer and the pumice dust, so light that it would rise two hundred feet in the air, a logging operation could be seen for 20 miles. Because of the choking dust, the cat skinners had to wear respirators.

The "60" cats did not lack power, though the horsepower was very low compared to later, bigger diesel cats. The difference was in the gears. "60" cats could handle big loads too, but they were slow. Still, as the ground was good for a cat operation, neither too steep nor rocky, the five cats shuttling back and forth, from the cutting areas to the landing, kept logs coming as fast as they could be loaded out. The chaser and the loaders dared not make a false move or they would get behind. The jammer puncher didn't get a break from his controls, and the barking, puffing and rattling from the machine, added to the loud exhausts of the cats as they came and went, kept an unabated roar around the landing. It was a welcome relief to everyone when the jammer whistle blew for lunch. The only time the whistle was heard was when it was time to start, or time to quit. No whistle punk was needed: hand signals were used by the loaders who were in full view of the puncher, as is the case in any loading operation.

During the lunch hour all those who had not dozed off from weariness amused themselves by feeding the big gray squirrels and watching an occasional curious deer approach from the uncut forest along one side of the track, quite unafraid, when everything was quiet, to look over the strange creatures who were doing such drastic things to their habitat. It was also a fine time for lying. The very susbtantial lunch sent out by the cook would be finished in ten minutes by the hungry crew. There would be a moment or two of idle conversation before someone would get down to the serious business of lying. If the liar happened to be Moffett, the skinny second loader, he would generally tell some tale about the biggest log he had ever

Showing ropes attached to the end-hooks. The loaders could pull end-hooks free and retrieve them with the ropes. *Courtesy Bob Lewis, Klamath Falls, Oregon.*

Trainload of logs at dump at Modoc Point. Owner W. E. Lamm on car. *Courtesy Favell Museum, Klamath Falls, Oregon.*

The light pumice dust almost hides the jammer as loaders fight to see and breathe as they go on with the gruelling work. *Courtesy Bob Lewis, Klamath Falls, Oregon.*

loaded, it would be twice as big as anything we had there; or he would tell how many loads he had loaded when he worked for Pokegama, it would be twice as many loads as we were now putting out, although the loaders and the puncher right there were putting out loads as fast as humans could move.

Then the ever present magpies would gather around, competing with the squirrels for tidbits of break, cake, ham, and fruit. The squirrels didn't eat the ham, which was a break for the magpies.

One man asked if we all knew magpies could talk. Seeing that he had the attention of those who were awake, he went on to say he was working on his roof one day and felt something pulling at his shoestring. He looked down to see that it was one of the magpies untying his shoe. Somebody observed that he was trying to con us. The man answered that honest to God he had kicked at the magpie and ordered him to quit that! He said the magpie jumped back and looked him in the eye, and said 'quit that'. Someone delcared the bird must have been somebody's pet, and they taught him to talk. We watched the squirrels and magpies for awhile in silence.

The jammer whistle blew and stiffly, reluctantly, the men got to their feet. Seconds later one cat, out in the timber, could be heard to bark into life, then another, and another. The head loader yelled for slack. The puncher crawled up on his seat and the afternoon turmoil began.

After a few weeks hooking behind the turn-out cat, and helping the choker setters, when we were caught up with the buckers, I heard one night that the cat hostler had been fired. I was only surprised that he hadn't *quit* a long time ago. Everybody knew he

"Sixty" cat crawling over log; if the way was blocked even by a small log, the cat would rear up and reaching the point of balance crash down with such force as to sometimes throw the "skinner" off. *Courtesy Bob Lewis, Klamath Falls, Oregon.*

had one awfully tough job. He had to gas up all six cats every night. The "60's" would burn about 50 gallons of gas a day, and the "30" would burn about 20. It all had to be hand-pumped out of a big tank on a sled. But that was only the beginning. One night a week the cat hostler serviced all transmissions, another night he did the final-drives. Another night he drained and changed the oil in all cats. Another night he changed all the filters. *Every* night he greased all track rollers and all other grease points. *Every* night he kept a fire in the jammer, and fired up a good head of steam in the morning. *Every* night he drained all the cats of water to prevent damage from freezing. *Every* morning he refilled all the cats with water, and started them up so they would be running, when the crew came in the crummy. But that was where the going got tough. Four of the cats wouldn't start when cold:

Falling ponderosa pine with 8-foot falling saw. *Courtesy Favell Museum, Klamath Falls, Oregon.*

Logging cat and big wheels. There is more than one way to start a cat. *Courtesy Bob Lewis, Klamath Falls, Oregon.*

dust had filtered into the impulses over a period of time so they would not function properly.

The cat doctor could have overhauled the gummed-up impulses so they could snap and spark. What he could have done, and what he did, were quite different. The hostler had employed every device that he could think of to start those stubborn cats. He got hold of some ether to prime them with. It fired all right, and he blew a jug (cylinder). That cat went into the shop, but still the "old man" didn't make the cat doctor straighten up. The hostler did have one cat that would start. He tried towing the others with that one. Being alone, he had to put the cat to be towed into high gear, then start off pulling it with the other. That worked fairly well except that he had to be nimble enough to stop the first cat, jump off and run back to the second cat, which had started, reach over the threshing tracks, and throw out the master clutch. He made it work. The only trouble was that one morning while he was running from the front cat to the back cat, he caught his toe under a root and fell down. The back cat churned up to the front cat and began to climb up on the back of it. The worst of it was the "old man" arrived at the landing just when that playful cat tried to climb the first one. So the hostler got fired. A miscarriage of justice, to say the least. The one who should have been fired was the lazy cat doctor.

The next morning Claude caught me at the crummy. He ordered me to stay in and sleep that day, saying I was going to change jobs. I asked doing what, and already knew the answer. He answered cryptically that I would be the cat hostler. I wanted to know what the matter was with my work right where I was. He said not a thing, that I was doing fine, that's why I was being promoted. I said I didn't think that was much of a promotion, so Claude pointed out that it paid more money and asked if I didn't need money. I couldn't argue with that, and besides he was the boss. Then there was another thing. I'd made a bad mistake bringing that guitar with me. There had been no rest for me in the evenings. I certainly was no Roy Clark on the guitar, and I suspect that my voice was not too superior, but the loggers liked it. There was no other entertainment in that camp. They liked the "bum" songs, the railroad songs, the folk songs, or anything else. Four or five of them could sing too, and we had a few numbers that were suitable for quartets. It wouldn't have been so bad if it hadn't been *every night.* Changing to night shift could change all this.

The only contact with civilization we had out there was the unfailing supply of gin. It wasn't real gin, it was just grain alcohol diluted with water and flavored with juniper berries. How it got into the camp in the quantities that it did I couldn't guess. I never bought any of it so I don't know who was selling it. Why it was tolerated in camp I can only conjecture. It dulled the urge to go to town, and kept the men reasonably contented. No man ever touched the stuff during working hours, and every man knew well that he had better be in shape to do his work. Many of the men lived in the family houses, and that too tended to keep employees in camp, but it's pretty hard to control what people do in their homes.

Anyway, the gin was there, and it was consumed in considerable quantities, particularly on weekends. Those weekends were hectic anyway. When men are used to burning up tremendous amounts of energy every day, and they get two days off with nothing to do, they will find outlets for their energy. I hadn't been in camp a week before one of the fellows grabbed me and insisted that we wrestle. There was little use telling him I didn't particularly want to wrestle, as we were already doing it. We threshed around on the ground, upending each other, each trying for a pin and neither knowing much about wrestling. Finally, half worn out, we collapsed on the ground, laughing with what breath we had left. Good-natured as he was with me, that same man whaled the daylights out of one of the loggers for a remark he didn't like. What a timber-falling man he was! Working on a big ponderosa with his partner, just when the back cut was nearly finished, he would yell around the tree to his partner to *saw, saw, saw*! Almost doubling their speed, most of the wood would be cut by the time the tree toppled. He never broke a tree, or had a "kickback".

The disaster of an exploding steam boiler. *Courtesy Pioneer Museum, Tillamook, Oregon.*

That night I started cat hostling. I made it through the night all right. When morning came I had a full head of steam in the jammer, and had managed to get all the cats running but one. I was worn out from reefing on starting bars. You started those cats by sticking a steel bar in a one-sided hole in the flywheel and giving a big pull. You continued that process until the cat started, if ever. If the cat started it was supposed to throw the bar free so you wouldn't get hurt. If the cat backfired, it could throw you across it with the bar, or ricochet the bar back on you, breaking your head.

That night I had done everything known to get the cats to start. I had opened the petcocks and primed each cat as I came to it, and fought until it started. One of the cats started right up, and the kitten *always* started. Three of the big ones I fought, and fought, and fought. One by one I finally got them started, all but one. It was still not running when the crew got there. One of the skinners got on that one, I tagged a line on it, and gave it a pull. The boss complained because not all the cats were running when the crew arrived. I told him he'd better make the cat doctor fix those impulses. But I knew the cat doctor wouldn't do it. He didn't care. He didn't have to heave on starting bars for an hour.

A month went by, and the boss found the cats running every morning when he got to the landing. He observed that I sure must have those cats figured, that it was the first time that year we'd gone for a month without the crew coming out to find two or three cats having to be towed. I grinned at him. I wasn't about to tell him how I'd solved the problem. The second night on the job while I was pumping gas, checking differentials, and forcing grease into the track bearings, with a hand Alemite gun, I had been thinking about the problem. I sure didn't want to go through the same battle every night as I had the last one. The cats had not cooled off yet when an idea hit me. As fast as I finished with a cat I pulled it up on a small hill nearby. I soon had the whole row of no-start cats up there. I drained them, and let them sit. After I had steam up in the jammer in the morning, I went after the cats. I tried coasting the first one and was delighted when it began to roll down the hill. That was the one thing I hadn't been sure of. When it got to clattering along pretty fast I threw in the master clutch. After a few revolutions the engine fired off. I repeated the process with all the stubborn cats, filling each with water the minute I reached the landing where the hose was. Inside of ten minutes I had all five cats and the kitten in a row beside the landing, talking to each other. I knew for certain that if the cat doctor knew what I was doing he would have screamed "to not run them cats at all without water in them". But I knew that I could have a cat to the landing and filling with water in less than three minutes, and that couldn't possibly hurt. Anyway, that was the way I did it, and would continue to do it until somebody overhauled those impulses.

About a week after the boss had told me how well he thought I was doing, an emergency arose. One night while I was working on the cats, the water in the jammer's boiler got a little low. When I jumped on her to feed the fire, I noticed that the water was barely

Jammer loading logs at Lamm Lumber Company's logging works. *Courtesy Favell Museum, Klamath Falls, Oregon.*

above the bottom of the water glass (a glass tube that shows the water level). I opened the water valve on the injector and gave it a little steam. I could tell by the sound that the injector didn't take. It was a cheap affair and the fireman had told me that it sometimes had to be coaxed to work. An injector had always seemed a minor miracle to me anyway. A small tube takes live steam from the crown of the boiler and leads it at a fractional angle into the stream of water that is directed to the cooler bottom of the boiler. It seems that the pressure inside the boiler would defeat that purpose, like trying to lift yourself by your bootstraps, but no. A good injector worked like crazy, putting 30 to 40 gallons of water into a boiler in a minute. However, this one was far from being good.

I shut the injector off, and tried again. This time I opened the water valve a little less, and the steam valve a little more. It sat there and chattered, but wouldn't start. I tried again, and again, and again, using various ratios of water to steam. The water had dropped out of sight in the glass. I tried some more. No good. After another half-hour, I knew I was in trouble. I pulled all the fire, but a bed of coals, and lit out down the track for camp. Since I was alone there, and had no transportation, I had no alternative but to leave everything and hike to camp.

As I made it down the track as fast as I could walk. I suffered torments of embarrassment and concern. I hated to wake the puncher, but it was three-thirty in the morning and I had no time to spare, I had to get him *now* and get that injector working or there was no way I could have a full head of steam in the boiler by starting time.

I heard a timber wolf up on the hill, and thought if he had friends with him he'd probably have a deer for breakfast unless a cougar beat him to it. I kept hitting the ties down the track, I probably covered that seven miles in record time. In camp, chagrined, I rapped on the door of the puncher's house. I rapped again. I heard a grunt and a stir inside, and the voice of Al, the puncher. He wanted to know who it was and what I wanted. I told him who I was, and asked him to come

At the landing: cat, big wheels, jammer and load. *Courtesy Bob Lewis, Klamath Falls, Oregon.*

to the door. There was more movement inside and the door opened. Needless to say he was surprised to see me there at that hour. I told him it was that injector, that I had tried for over an hour to get it to start, and just couldn't, and when I left there the water was out of sight in the glass, and had been for a long time. Al exploded, cursing the injector, but neither of us wanted to arouse half the camp. Al got dressed and we went down to the shop area and got into a utility truck, heading for the woods. Pulling up to the jammer, Al jumped on and looked at the steam pressure.

He told me to get the thing fired up hot, just as fast as I could. I got some small wood in on top of the bed of coals and added some pitchy pieces on top. The fire began to roar and it wasn't long before we had steam coming up *fast, too* fast.That told us both that there was very little water in the bottom of the boiler. I remembered about a time when a boiler blew up, wiping out everybody and everything for several yards around. Al worked with the injector. I could see the perspiration standing in beads on his forehead. Again, and again, and again he tried to get the injector to take. I think he was just on the verge of telling me to pull the fire, because he had turned to me with a look of hopelessness on his face, when suddenly the injector got quiet. We could hear the sound of whispering water in the tube as it sluiced into the boiler. the injector was working!

Al gave a sigh and dropped down onto a pile of wood. He wasn't alone. He looked at his watch, noting that it was quarter to six. He thought he may as well just stay there. I argued that he had time to go to camp and get breakfast before the crew came out, and that they might need the truck. I didn't think he'd go, but he did. I had urged him to go because I had to get all those cats started, and I sure didn't want him to see how I did it.

CHAPTER 12
THE WAY IT WAS

Over the Fourth of July a couple of extra days off were scheduled to allow for work on the Paul Bunyan wheels and other repairs, so we would have five days. The last thing I did before I left camp was to walk one of the "60" cats the seven miles into camp to the shop. I hoped that by the time I got back all those infuriating impulses would be overhauled.

I got on one of Lamm's trains of logs and rode in to Modoc Point and caught a truck to Klamath Falls. On the way in we stopped at one of the section gang stations, where a young man from Minnesota got on the train. Riding in to Modoc Point, the young fellow asked me where I was going. I told him I was going to Salem for a couple of days. He said he was going to Portland to look around. Cliff, my new acquaintance, said he would like us to travel together, to ride the freight rather than waste money on train or bus fares. (I hadn't intended to.) He said we could catch the Los Angeles to Seattle Manifest in Klamath Falls. I agreed, as I was going to need to get to Salem with as much money as possible.

In due time we found ourselves in the hobo jungle at the end of the old railroad overpass at Klamath Falls. Cliff and I had agreed before that we would keep the possession of money strictly to ourselves. We had had something to eat uptown before we crossed over to the jungle. While crossing the overpass, Cliff told me a story: the last time he had been in Klamath Falls he was crossing the overpass to go uptown for something to eat. Somehow some of those hobos got the idea he had some money on him. Right in the middle of the bridge, a man in front of him turned around and waited for him to catch up. He asked Cliff if he had a match. He started to feel in a pocket to get him a match. All at once he heard the faintest kind of sound behind him and it hit him like a flash. The one asking for the match was setting him up. He hit the man in front of him as hard as he could and knocked him down. He spun around just in time to see a man with a club ready to break his head. He hit him, too, knocked him flat. Then he ran like the devil. I said we'd have to play it cool. I had never seen times like they were right then and guessed some of those men would kill for a dollar. Cliff amended my comment. He said they might do it for two bits. He said starvation was so bad because when a train pulled in to one of the California towns and the men tried to get off, there were numerous police there that clubbed them right back on the train again. It seemed they couldn't even go into a town and try to beg a little something to eat.

In the hobo jungle I began to see just how bad things were for the down and out, unemployed Americans who had to join the 'bos, scrounging for a little food wherever they could. Stealing a few carrots, or a head of cabbage or any other vegetables they could find in the fields nearest the tracks. All tried to come up with something; most failed. Still, every night there was a stew brewing in an old wash boiler somebody found. Water was the main ingredient in that so-called stew. Bits of meat, and what not, came from the garbage cans at the alley door of every restaurant to be found. The diet was a starvation one. An old tin can served as a soup cup. Cliff and I were afraid not to eat any of the mixture. Somebody might suspect that we had some money if we didn't.

Even though I had seen the Southern Pacific trains go through town loaded with men, I was shocked to see the number of men in that jungle, most of whom would jump the Manifest that night. I knew that in every railroad town in the United States there were that many men or more in a nearby hobo jungle, to say nothing of the millions who were constantly on the freights. Some were in the remnants of

A hobo jungle, Klamath Falls, in depression years. *Courtesy Brown Brothers, Sterling, Pennsylvania.*

what had been decent clothes. Some wore old garments obviously picked up from some trash pile or city dump. There were men who wore clothes far too big and there were others squeezed into old coats that had sleeves too short for their arms.

That evening we listened to bitter stories of lost farms, bankrupt merchants, skilled workers who could find nothing, and men who hadn't seen their families in months. Displaced persons; shells of their former selves; miserable, disillusioned and pitiful.

Hardened as I had imagined I was, I couldn't help feeling a wave of pity sweep through my soul as I looked into the depths of hopelessness in the eyes of so many of the men clustered around the little fire. The boiler of "jungle stew" boiled constantly as men came and went.

About ten o'clock a distant whispering rumble disturbed the night. It swelled to a roar. The Los Angeles Manifest was pulling in. The big 2-8-8-2 Mallet type engine, its headlight searching out the grade a mile ahead, came groaning to a stop in front of its mile-long train of produce and other freight from all over southern California. Hobos crawled off and headed for the jungle where they would lay over and get something to eat, bad as it was. Other 'bos made it for the train to find the empties and climb in.

Cliff and I found a car and jumped aboard. Brakemen came and went, swinging their lanterns and paying no attention to any of us. In those depression days men were not kicked off trains, as they used to be, so long as they caused the train crew no trouble.

Mallet engine pulling a train across California. *Courtesy Golden West Books, San Marino, California.*

Soon the big Mallet began to surge into its load. The shock wave travelled from the front to the rear of the train, and the thundering roll of the tightening drawheads ran the length of the train as the engine took up the slack and the iron snake began moving, leaving Klamath Falls behind and rumbling along the contour of Klamath Lake, heading north. We gained speed, passing Algoma and Modoc Point. The freight began to roll faster. The floor of our boxcar, empty except for human bodies, developed a bouncing motion. I knew this was caused by crossing the rail joints at a fast rate. We started to settle down for the night. There was a quantity of shipping paper in the car. We put sheets of it under and over us. To my surprise, the paper made a remarkably warm cover. Hard as the floor was, we were soon asleep.

The train began to slow, the bouncing stopped. At last we eased into a section of double track and stopped. Someone mumbled that we were waiting for a south-bound. In spite of the hard floor, I was not too uncomfortable. I went back to sleep. I was awakened by a hand on my leg by my pocket. It was Cliff. He explained that he was just getting up to go to the door. I wondered if he was really trying to roll me, knowing I would blame one of the others. My money was not where it could be found that easily, and it didn't happen again.

We must have been on the double track quite awhile before the darkness began to be diluted by a faint, flickering light. The darkness melted as the light became bright, then a glaring headlight of a fast passenger pierced the black night and the train roared past the half-open door of our boxcar. glimpses of coach windows with the blinds half pulled, the fully lighted diner and dark sleepers were visible for split seconds as the train flashed by us and the sound faded into the distance. The fast train rolled towards Klamath Falls.

A typical hobo riding the rods on a freight train. *Courtesy Brown Brothers, Sterling, Pennsylvania.*

The freight shuddered again and I felt her surge. In the still of the night I clearly heard the authoritative bark of the Mallet as she heaved the long train into action again. I went back to sleep. I didn't waken again until I became aware of quiet. We had stopped again. Soft light, as from nearby street lights or railroad yard lights filtered through the night. Men stirred and a few sat up. One man asked where we were if anyone might care to answer. Someone said it was some place up in the mountains, with a spot or two of snow on the ground. He thought we were waiting for another train to pass. A brakeman walked by with his lantern, his feet crunching the cinders. Our eyes adjusted to the dim light, we could see those near the door. One sleazy character sat up and grumbled that if it wasn't for the rich farmers we wouldn't be in that shape. I looked at him and knew he wasn't one of the good men who lost his job and his family, and his chance in life. Even in the dim light I could see "professional hobo" written in his face. He was about thirty, dirty and evil.

He continued profanely that he had never seen a farmer in *his* life that didn't have it made, with plenty to eat, a good bed to sleep in, and plenty of money in the bank. A man in his sixties sat up tensely, then jumped to his feet. He cautioned the sleazy one not to run off at the mouth about things he didn't know anything about. He said he was a farmer, and had lost his farm in that depression as a lot *more* had done. He resented any fool trying to persuade anyone that the whole depression was the fault of the farmers. The sleazy man got to his feet. He said he didn't like being called a liar. The older man didn't see why that should bother him, since that was what he was.

The younger man seemed to turn a yellow color in the dim light. He was going into his pocket for a knife. I knew that if nobody stopped this, the older man was going to get carved up, if not killed. I roared with all the authority I could get into my voice, as I jumped to my feet, that if they were going to fight, they get off the train and do their fighting out there on the ground, or they would get us *all* kicked off the train.

The older man had seen the hobo going into his pocket, and he saw, as others of us did, the knife that appeared briefly, to be dropped back into the filthy pocket as the scum backed off and slouched to the other end of the car. The farmer threw me a look of gratitude and nodded.

When the sun came out and some of us had made our way on top of a pile of crates loaded on a flat car, the old gentleman eased over to me. I clearly recall what he said. "I got two sandwiches here that I got yesterday. I want you t' have one of 'em." Embarrassed, and a little ashamed because I had no need of this kindly offered help, I said I didn't want it, but I was grateful for the offer. He insisted that it was good food and that I had to have food. I didn't dare say that I wasn't broke. Ashamed as I was, I had to let the old man believe that I only wanted him to keep his food for himself. Offended with me because I wouldn't let him show his gratitude to me for saving his life, the old man fell into a moody streak. I knew he was thinking troubled thoughts about his old home, and probably his family some place, I thought it best to leave him alone.

That morning we had passed through Oak Ridge, and come to Eugene, where I saw a man swing off the east side of the train while it was moving too fast, but didn't see him reach the sidewalk. We reached Salem and I dropped off the train, and walked four miles to my house. I was glad to be home, even for a couple of days. But I was saddened to read in the *Statesman* next day that a man had been killed in Eugene the day before when he had jumped from a moving train.

There were groceries to buy, there were things to be done around the place, and there were friends to see. Most of all, there was my family. The kids had-grown noticeably even in the brief time I had been away. Too soon it was time to start back to camp. My wife and her brothers tried to talk me out of going back on the freight, but I had learned the value of a dollar in those lean years and I refused to spend the price of a ticket.

It was raining when my wife and her brother dropped me off at the freight yard in Salem. My wife was almost in tears. The through freight pulled through Salem and I swung aboard the back end of a lumber car. Overhanging lumber protected me from the wind and most of the rain. It was late in the evening when I boarded, so after a number of stops it was getting daylight before we had reached the summit of the Cascades. Somewhere, either at Eugene or at Oakridge, we had picked up an extra Mallet. The reason for the extra "hog" was not that the big engine couldn't pull the train up those mountains, but that it might pull the middle of the train off the tracks on the switchbacks. The extra Mallet was put into the middle of the train to ease the strain.

On top of the lumber cars, as the sun came up and it began to warm up, I was surprised to again run into Cliff, on his way back to Lamm's too. He had picked up a friend in Portland. The new man had only a sweater to keep out the wind. Lying on top of a car of lumber, we were drying out. The three of us were lying there together, I had a big heavy overcoat. Cliff

Believed to be small early Shay, left. Chained loads of pine. *Courtesy Favell Museum, Klamath Falls, Oregon.*

Hobos on top of freight car in depression times. *Courtesy Brown Brothers, Sterling, Pennsylvania.*

had a mackinaw. We came to one of the long tunnels. We failed to realize the problem until it came. The tunnel was full of acrid smoke from the two locomotives.

Cover your face, bury your face in your coat, I heard Cliff yell. The little guy with the sweater was lying between us. He started to jump up. He would have bashed his brains out on the roof of the tunnel. He was suffocating with the heavy oil smoke. Cliff and I grabbed him from both sides and held him down in spite of his struggles. While pinning him down we kept our own faces buried in our coats. It was pretty horrible. When we finally came out the upper end of the tunnel we could breathe! We turned our prisoner loose. He was only half conscious. He sat up and swayed, his eyes glazed, and two streams of black came down from his nostrils. I think he came close to fatal asphyxiation in the tunnel; with only the sweater to filter the smoke he had no protection.

The rest of the run to Klamath Falls was without incident, but the sight of those hundreds of men riding the freights, hoping for some kind of relief from their hunger and suffering made me wonder what our country had come to. The trainload of men heading south would pass, every couple of hours, a train crawling with men heading north. It seemed so futile and ridiculous.

Back at Lamm's camp, I felt more thankful than ever that I was not one of those men on the road. I knew something had to happen in this country to reduce the terrible rate of unemployment, but I didn't know any answers.

The jammer had a new injector when I got back on the job. The cats didn't start any better than before. Pine timber continued to pour into the landing. Every man put all he had into log production. The long train of logs left for Modoc Point every day. The summer had gone and fall was in the air. It froze hard every night but the days were still comfortable.

I worked until a shutdown came, late in the fall, then returned to Salem on a freight train. The men riding the freights looked worse than I remembered them before, and there were more of them.

Cars of logs being dumped, Modoc Point. Two ponderosa logs floating in foreground. *Courtesy Favell Museum, Klamath Falls, Oregon.*

CHAPTER 13
HARD DRIVING

The following spring I rented my Salem place and moved my family to Tillamook where I had a job as caretaker for the old Hammond Tillamook headquarters camp at Idaville. There I boarded up the windows and did repair work on the old storage warehouse, and the hotel. The work was temporary but I had the use of the old superintendent's house for as long as I wanted it.

It was not long before I got an opportunity to take a log hauling contract. Between my persuasion and the fact of the contract, I got a truck that was only a year old and a big single-axle trailer.

The first thing I had to haul was the gas donkey that would be used for yarding. It was in pasture land north of Tillamook where it had been yarding and piling stumps. We got the donkey pulled up onto a couple of logs and slipped it sidewise onto the truck by using the donkey's power and a couple of blocks. I didn't think I could pull the heavy load across the pasture land to the highway, but I made it.

South of town, at Simmons Creek, the right-of-way to the timber began. I made it over the soft, new road to the foot of the hill where the timber was and we unloaded the donkey. We began hanging blocks on stumps and pulling the donkey up the hill. We rigged the tree at the same time we graded the road up the hill.

Finally the tree was ready. We had thousands of feet cut and ready to log. I got the truck up the hill and up to the brow log. The puncher already had logs pulled to the landing. I acted as loader and we piled on a load that I learned later was heavier than I should have had for the first load. The front wheels almost left the ground with the pull required to start the load. Down the hill I went, and I felt a surge of joy. For the first time I was doing something I had wanted to do so much. Those first logs were soon in Tillamook Slough.

The allowable weight on the highway was 14,400 pounds per axle on six-inch rubber and dual wheels on truck and trailer. There were two schools of thought amongst log haulers at that time. Some liked a big, heavy truck that would last for years. Others liked a Chevrolet or Ford which they hoped to be able to replace in a year or two. The light trucks could get over the scales with as much wood as the heavy trucks, even with some difference in tire size, because they would weigh one or two tons less. I had a Chevrolet, not by choice, at that time, but because it was all I could get with extremely limited finances.

I was hauling alone on that first job. Later I found I could move as many thousand feet of timber in a day as the big heavy trucks could.

As the road on the hill hardened, I discovered that my load wanted to run down the hill too fast; I had no trailer brakes. We devised the idea of stringing an old mainline around a stump at the top of the hill. With an eye in each end of the line and the bight around the stump, it was simple to put one eye of the line on the reach pin before starting down the hill. In order for the truck to go down the hill, it had to pull that old mainline *up* the hill, plus offsetting the drag and the friction of the line around the stump. The next time down all I had to do was hook onto the opposite end of the line and the exact same friction and weight held me again. By the time I had finished that first contract I had a new trailer, with vacuum brakes.

The day after I finished that job, I got a call from the man who had been hauling the planks for the logging roads of the Coats Logging Company. He wanted to know if I would consider putting my truck on a plank haul. I said I would. He told me to go to the Coats machine shop up Bewly Creek. I had worked for Coats years before, of course, while they were still railroad logging and I knew where the old headquarters camp was. I drove up there the next morning and stopped in front of the machine shop.

I told the blacksmith that I was told to come there with my truck but I didn't know why. He said I must be the new road-plank hauler, and that they had to make

1933 Chevrolet truck hauling gas donkey.

a push bar for my truck. I asked why a push bar, that I didn't get it. He said I would understand the first time I tried to back a truck and trailer up a two-plank road. I said no more.

After awhile the blacksmith had finished the masterpiece. They loaded it onto my "jockey box" and I drove to the old Coats side west of Tillamook. I got to the road, jackknifed the truck around and backed up the plank road to where it was being taken up. The crew was not there, but they soon arrived, and the boss looked at me. He asked me how in the world I got my rig back in there. I told him I backed in, of course. He wanted to know how I could do that without anybody to steer it, that it was a half-mile. I said I had supposed that I would be expected to be able to do that myself. Olds said the other driver never could.

He explained that the contraption the blacksmith had made was a bar to connect the truck and trailer so the trailer reach could be unhooked and a man could stand on the back of the truck holding it and steering the trailer so the truck driver would only have the truck to steer. I never used the thing, and got it off the truck the first chance I had.

This was my third experience in trucking. I had once had a Ford Model TT with which I hauled wood, coal, and alder logs. Then I had worked with a friend near Salem who owned the most amazing Model TT I ever saw. It had two transmissions and a two-speed rear end. It had fourteen speeds ahead and ten in reverse. It was my job to drive the truck, hauling wood in a variety of lengths for the old prune driers, schools, and private homes. Obviously we couldn't afford to put roads in to every tree we cut. The most we did was take a little off a high spot with a shovel and throw it into a low spot. For the most part, we just drove the truck around in the woods from one woodpile to another. Coming out, the load would sway and weave as the truck struggled over the rough ground. Frank, the man I worked for, delighted in telling people that I could go through the woods backwards with that truck faster than most people could go forward.

The operation I had just begun, at Tillamook, used a plank loader. It had a gasoline engine and was mounted on a four-wheel trailer with a conveyor belt that could be adjusted for pitch to suit the height of the load as it increased. The feeders below dragged the planks onto the lower end of the belt with two-man tongs. The loaders on the truck worked with picaroons, dragging the planks into position to build the ever-heightening load. The conveyor belt had to be kept in adjustment so the men could do their work with reasonable ease. A load could be put on quickly by this method. My first load was soon on. I tightened it down with chain binders, and started for the highway. While I was gone the crew would tear up more planks and stack them close to the loader so they would be ready when I got back.

I was soon pulling into the plank road to the new logging side at Yellow Fir. A quarter-mile in, I came to the dispatcher's shack. He gave me clearance to proceed to the next turnout where there was a telephone to use for any further clearance. Just a bit above the dispatcher's shack there was a steep hill. I started up and found that the truck had about all it could pull, even with its low ratio differential. After that first load I could tell how much I dared to put on and still expect to make the hill.

The hill was long as well as steep. After another clearance, I arrived at the landing where I passed a truck just pulling out with a load. As customary, there was a wide planked area there so trucks could turn around, and pass each other. On beyond the landing, where the road was being extended, I saw the second reason the former plank hauler had given up his job (the first reason was the hill). The crew was starting to plank a truck road over an old railroad trestle that was long, curved and very high. I knew that I was going to have to back off that trestle, after unloading, each trip until the road was laid beyond it. In some forest fire, years before, the trestle had sustained some fire damage to its piling below. Coats' engineers had checked it over and decided it was strong enough to support log trucks, but some of the truckers had their doubts.

I discovered that there was a benefit to the plank hauling job that I had not anticipated. I was coming

Going up empty, stopping at phone in box by cable spool, to call dispatcher. Coats Logging Company, near Yellow Fir.

back through the landing after being unloaded. There was not another truck there. Because I was hauling 20-foot planks, my semi was coupled short. That day there happened to be a big, short log on the landing. Big diameter of course means high board-foot content. The foreman flagged me down and asked if I would want to take a load on my way back. I said sure. This was what truckers dreamed about, taking a payload each way. The logs at that time were going into the slough at Tillamook and I had to pass near there anyway. Soon it became standard practice that if there were short logs on the landing when I came through, I got them, Nine times out of ten they were fine logs that really paid to haul. Logs are generally cut short because they are too big to yard and load in long lengths. There were times, of course, when there were only long logs on the landing, and I went back empty, but I didn't worry, I was getting my share of scale every month, and then some.

The road project progressed. I got so used to the trestle that I didn't mind backing across it to turn around after unloading. Soon the trestle was all planked and the road around the hill beyond. When the railroad had operated in there, the value of stumpage was low. Patches of timber were often left only because they were a little hard to get. Then when the fire damage occurred, the company had left the remaining timber to log other holdings. Later when stumpage became valuable, the trestles would no longer support trains, and the rails had long ago been removed, so back they went after the timber, building truck roads, and using the old grades wherever they could.

It had to happen sooner or later. I had consistently hauled every pound of planks my truck would handle up the Yellow Fir hill. Watching the loaders, I would stop them when I knew I had enough on board. One day I was pulling the hill with a load that was really a

First landing up the hill at Coats Logging Company. Loading donkey in front, yarder in back to the right (note dangling tongs).

Side view of the high trestle once used for railroad, being planked for trucks. Coats Logging Company, near Yellow Fir.

Hilltop view of the high trestle before it was converted to truck road. Coats Logging Company, near Yellow Fir.

#2 1935 Chevrolet with 4500 feet of spruce, Tillamook. Coats Logging Company.

On the high trestle. Coats Logging Company, near Yellow Fir.

1935 Chevrolet just below the big trestle at Yellow Fir side. (Coats). Truck road where railroad once was, approaching top of big hill.

line case. If my truck didn't falter once, I should make it. Just as I was approaching the stiffest part of the hill, I saw four men standing by the road. There were some blocks and other heavy gear on the ground beside them. It didn't dawn on me until it was too late what their intentions were. As I started to pull past the men, I realized to my horror that they were throwing all that heavy stuff on my load. What was even worse, they were climbing on themselves! They were adding a total of well over a thousand pounds to a load that didn't need ten! One of the men was the foreman, who should have known better.

I started screaming for them not to put the stuff on there, and to get off themselves. The truck pulled down, shuddered and quit. The motor even started spinning backward. The vacuum brakes were useless without the motor running, and I couldn't hold half that load with the truck brakes. I was going backward down the hill with a canyon dropping away on the left. Suddenly, with no choice in the matter, I had the responsibility for four lives besides my own. I'll admit that at that moment I could have killed those brush apes for the stupid thing they had done, but I wanted to do it with a club, not this way. There was only one thing I could do. I did it just as quickly and violently as I could. I spun the steering wheel and threw the truck off the road in a jackknife. The truck turned crosswise of the road and poured off planks, men and gear all over the road. No one was hurt. The following is reasonably close to what was said. I yelled at the foreman if he couldn't see the truck was "pullin' all it could and ain't you got any sense at all?" He replied that if he'd "spit right quick maybe ya' coulda made it", hoping to appear master of the situation to his men.

The foreman was really decent though. He sent a man up the hill for more men. I got the truck squared around so it could be reloaded, and he saw that his men loaded all the planks I could start on that hill, which was considerably less than I had before. Best of all, he chose to overlook the bawling out I had given him.

That night, telling about the incident to a mechanic friend, I asked why we couldn't build a tank with ground seat valves that would retain vacuum, the same way tanks hold air? We started working on the project in the evenings. Before long I had the only vacuum-retainer tank on a truck in that part of the country, or maybe any part. That thing worked beautifully. The normal running of the motor kept the tank drained to almost perfect vacuum all the time. When I needed brakes I pulled the Bendix valve, the same as before, but the valve worked between the pre-vacuumized tank and brake boosters, giving an instant full application of brakes, if needed, without causing the faltering of the motor that we were all used to with the old direct system.

Once after that I had to stop while pulling a hill, to make a minor repair. I set the brakes, took my time about the work, started the truck again and continued up the hill. The retainer tank *still* had plenty of vacuum.

We had continued the plank road around the hill and had built a section up a steep hill to reach a stand of good, old, fir and spruce. That was a tough stretch to build, and to haul the planks up, but it was finally completed, including the grading of a turn-around for the trucks. At the new landing, the spar tree was already being rigged. In a day or two the lower operation was finished and the upper one was ready.

One morning all the trucks and the crew crossed the trestle and headed for the upper landing. I had become used to the trestle while we planked it, and the road above. The other drivers crossed the long span for the first time that morning. At the landing there was a discussion over who would take the first load down. Alex Gigioux was there with his big Federal, Frank Reed with his Mack, Bruce Elliot with his Reo. Pop Hughes had two trucks on the job, a big Jimmy and a small Jimmy (GMC). Cecil Hughes also had a Jimmy. There were some other trucks too; one, belonging to Louis Welch, was the heaviest rig of all. It was a big Cornbinder (International), and Louis had built his own trailer out of parts of trucks fastened together, not by welding but by hundreds of pounds of channel iron, angle iron, box steel, I-beams, nuts and bolts. It was the heaviest trailer I had ever seen.

In the end, it was decided that one of the lighter trucks should test the trestle first, though it really didn't make any difference, the gross weight would be about the same anyway. None of the drivers of the other light rigs were bursting with eagerness to be first. I was used to the trestle, and had already hauled some fairly heavy loads across it, so I settled the question by backing my outfit under the spar tree. I had just put in a long reach because plank hauling had been discontinued until another new road had to be built. The head loader asked how much "wood" I wanted on the truck, suggesting that perhaps it should be a light load. I said if we were going to test the trestle, we might as well test it, and to wood her down. The loaders had two logs just right for a bunk load. They loaded them. The loader asked if I wanted more, and I said I could take a peaker, if it was not too big. The loaders jumped across some closer logs and threw their tongs on a good-sized log which they put

High on springboards, ready to drop a spruce. Tillamook County, near Yellow Fir. *Courtesy Lila Boge, Tillamook, Oregon.*

A medium but perfect spruce tree on its way to the ground, cut high because of logs and debris around butt.

on top. I hadn't wanted quite that much but I wouldn't say so. I knew, as I pulled away from the landing and dropped over the first hill that every man at the landing went to the brow of the hill, where they could see the trestle, to watch until I had crossed. I must confess some misgivings. The load I had aboard would weigh nearly twice as much as the loads of planks I had hauled across the trestle.

Down the first hill, carefully braking the trailer more than the truck, I rounded the head of a draw to the right, then swung left around the toe of a ridge where for a couple of minutes I was out of sight from the top. The next time the interested viewers on top could see my truck, I was starting across the trestle. Just to cheer them up, I let her roll out onto the trestle nice and fast and was completely across in about a minute. I could hear that old trestle creak, and I could feel it sway under me as I crossed, but it was still standing.

After I had dumped my load into the new booming ground in the Tillamook River, and was returning to the woods, I began passing other trucks coming out with their loads.

All the drivers quickly got used to the height, the narrowness and creaking of the trestle and would cross it at twenty miles an hour. Others, coming into the works for the first time, didn't like it. One day the Columbia River scaler rode up to the woods with me. Coming upon the trestle unexpectedly, he nearly went into shock. Leaning to the center of the truck, he turned pale. After he had finished his inspection of the cut timber, he was ready to go back with me. He told me, when we got to that trestle, that he wanted out, he was not about to cross that thing riding a loaded truck. I respected his feelings and stopped just before we hit the trestle. He got out and I pulled on across and stopped. I got out of the truck and looked back to see how the scaler was doing. There he was a little way out on the trestle, down on all fours, crawling and shaking, while I waited. Once in the truck, white and trembling, he said nothing all the way to the dump. I think he was incapable of talking. As long as we worked at Yellow Fir, he never came back to the woods again.

I had needed a new truck for some time. Overloading the way we all did, I would lose a rear axle once in awhile. I finally adopted the practice of carrying a spare with me along with the tools to install it. Also, the frame had developed a chronic tendency to sag. Every Saturday I would sling a big timber under the truck with chains. Then I would move along with a hydraulic jack pushing up on the frame at the right points. I didn't want to "fishplate" it because of the danger of the frame breaking at the ends of the fishplates.

My job seemed permanent, so I ordered a new truck and a new trailer. I kept that wonderful vacuum retainer tank and put it on the new truck. The intake manifold had to be drilled and tapped to install the nipple for the vacuum hose. In the middle of the process, it became necessary for me to leave town. I left the truck at the Tillamook garage to have the remaining work done. When I returned I was told that my truck was all ready to go.

In the morning I headed the new truck for Yellow Fir, with the new trailer behind. Trailers were not loaded on the trucks then, as they were later, for running empty. So on the trip to the woods that morning, I had used the trailer brakes lightly a few times before I took on my load at the landing.

Leaving the landing I had the first steep hill to drop down, but the vacuum retainer tank had enough vacuum to hold the trailer all the way down that first steep pitch. I proceeded across the trestle unaware that anything was wrong with my brakes. Just at the top of the *long* hill there was a telephone stop. I pulled the valve for the trailer brakes, nothing happened. I left the lever in the "full brake" position and got on the truck brakes hard. It wasn't enough. I couldn't get stopped to telephone on the grade. I managed to gear down to second gear before I broke over the big hill, but couldn't get any lower because the truck was moving too fast. I was on the main hill, trying to push the brake pedal through the floor. At the same time I was pulling back on the hand brake with all the strength I had. In spite of everything I could do I was committed. I was on my way down that long hill, and worse I had not called the dispatcher.

Trying to hold the truck on the road with my left hand while I strained on the hand brake with my right, I came down that plank road at an unbelievable speed. The weight of the big load of logs was pushing the motor to such a speed that it rose in pitch to an angry scream. I felt sure the motor would fly apart from being turned five or six times the speed it was engineered to run.

The road was the "two plank" type allowing 24 inches on each side to accommodate dual wheels. That meant that the front wheels which aligned with the inner rear ones, had only a foot tolerance. It was more than I could manage with one hand. I lost a front wheel onto the cross-ties. I heaved it back on only to lose the other one on the other side. I gave up on the hand brake and concentrated on getting both front

Chevrolet being loaded with 40-footers. Yellow Fir side, Tillamook. Coats Lumber Company. #2 just right of spar tree.

"Jimmy" approaching Highway 101. *Courtesy Joe Bell, Tillamook, Oregon.*

Loaded down, approaching the dump. Coats Logging Company, Tillamook. *Courtesy Joe Bell, Tillamook, Oregon.*

Dumping logs into river. Courtesy of William Jones.

wheels on the planks and keeping them there. Words fail to describe the high-pitched scream of the truck. The dispatcher heard it in his shack. No one had called him, so he knew it was a runaway. Fearing every second that I would see a truck coming at me up the hill, I rode the flying truck right on by the dispatcher's shack. He didn't help a bit when he got so excited he left his telephone and ran out, cursing and waving his arms, and ordering me to stop. Seeing his excitement, I felt positive that another truck was on its way in. I hoped that if I did meet one I could ditch my truck some way that would not kill us both. The county road came in sight, and I sweated and prayed, just a few hundred more feet. I rolled out onto the county road just in time to clear an oncoming truck by inches.

Everyone dislikes the anticlimactic, but it seems necessary to explain the loss of trailer brakes. The mechanic who hooked up my vacuum system found that he had tapped the manifold for too small a nipple to fit the hose. Instead of drilling a larger hole and tapping for the right-size nipple, he wrapped the smaller nipple with friction tape and clamped the hose on it. The first time the brakes were used hard the motor had to revacuumize the tank. With the motor turning fast, the vacuum at the manifold was at maximum, and the friction tape couldn't stand the suction. It came out from under the clamped hose, and effectively plugged the manifold nipple. The result was simple, and disastrous. There wasn't a sign of brakes to the trailer after its first hard application. For some unaccountable reason, the truck motor showed no bad effects from the abuse, and I punished it hard for two years, never losing a bearing or valve. But, from that day on no one worked on my brake system but me.

CHAPTER 14
MEN

Abe was one of a kind. He was lean and slightly taller than average. He may have been of Scandinavian descent; the only clue to that is that when he would stop at the dispatcher's shack, if the dispatcher's wife happened to have some cake set out for the drivers who might happen in. Abe would shout "Keken!" Abe was jovial and friendly with everyone but close to no one. The most notable physical attribute about Abe was his face. I always called him *Honest Abe* because of his homeliness. His eyes were blue, wide apart, and high. His cheeks were hollow and his chin square. His nose appeared to have been broken several times, it wandered from its point of origin between his eyes in several directions on its way to its termination above the upper lip.

Abe was a wild man. He drove the big GMC for "Pop" Hughes. He hauled on a percentage basis, so he was as log-hungry as any of us. Yellow Fir was not the best place for Abe to haul, because of the long hill. He always insisted on his truck being "wooded down". The problem with the big Jimmy was that it had inadequate brakes. That didn't stop Abe. He had a big single-axle trailer, so he could carry a lot of wood, and because his truck weight was not excessive he could get over the state scales with a lot. Permissible weight per axle was partly governed by the size of the tires on any given axle. Abe's trailer had ten-inch rubber, four inches bigger than most.

Manhood has all too frequently, and for too long, been equated with ability in the art of fisticuffs. Nothing could be less justified. Honest Abe took chances every day that would have struck stark, clammy fear to the heart of Jack Dempsy. Day after day, trip after trip, Abe broke over the big hill with five to six thousand feet of heavy timber, riding his rig inexorably, relentlessly toward the bottom. Nothing Abe could do, short of ditching, was about to stop the truck and its 15-ton cargo from *reaching* the bottom. It scared the living wits out of me when I went down that hill *once* out of control. Abe did it every day. If he felt any fear, he was an artist at acting, because he never showed it.

It is true, Abe's truck was not so completely out of control as mine had been. There was no danger of his damaging the engine. He had *some* brakes, but not enough to control the loads he was hauling. Many times I have waited by the dispatcher's shack when Abe called in from the traffic control phone just at the top of the long hill. In a few seconds we would hear the truck begin to pick up speed on the hill. Abe would not take the time to come down in compound (the lowest gear). Like the rest, he wanted to get as much timber in the water as he could every day. He didn't think he could get enough by crawling down every hill in compound.

The farther down the hill Abe progressed, the higher in pitch was the sound of his engine. His exhaust had a tendency to crackle like the sound of a machine gun. When the load was pushing hardest on the steepest part of the hill, even from where we waited half a mile away, the big Jimmy sounded like the approach of a dive bomber. Somehow Abe always managed to get stopped at the dispatcher's shack for clearance to go on.

There was one time, while we were still hauling from Yellow Fir, that Abe should have stopped but couldn't. The highway scales were open that day, and Abe came wheeling around a curve to see a line of four or five trucks pulled off on the ramp to the scales. There was no way in the world he could get his hog stopped, and he was overloaded by about a ton and a half anyway.

It didn't take Abe half a second to decide what to do. He dropped the two-speed axle into low range and tried to shove the throttle through the floor. The big Jimmy sang out like a banshee and ripped on by the line of trucks, the scale, and the weighing officer. The State cop demanded to know who it was as he raced out of the scale house in a rage. He demanded over and over again, but got replies like "I don't know", "I never noticed who it was", and "Looked like he had a light load anyway". Everybody lied because they knew who it was, but there wasn't a man there who was about to tell that cop. The cop had the trucks

Approaching the dump in Tillamook Slough. Big logs were common in those days. *Courtesy Joe Bell, Tillamook, Oregon.*

to weigh and he was alone at the station. There was no way he could capture Abe, so he gave up in disgust, threatening that he would get him next time.

We had finished logging the Yellow Fir side, and I had gone back to hauling planks while we laid the first quarter-mile of the Sand Lake road. Then two donkeys were moved in by truck, and the spar tree was rigged. Falling and bucking had been going on there for two weeks before the tree was ready. It was spring and the gravel roads were spongy, road shoulders were soft.

One day Elliot pushed the nose of his big Reo toward Sand Lake Hill with the first load out of that side, and the logging began furiously. The haul from Sand Lake was much longer than it had been from Yellow Fir. Truckers seemed to push their rigs even harder than before. Fixed expenses, such as truck payments, insurance, PUC permits and living costs remained static each month. The number of loads hauled per month was the thing that could make the difference between just meeting expenses, and showing a profit. By pushing, it was possible to bring five loads a day in from Sand Lake to the Tillamook River. Every man tried to get that fifth load.

True to form, Abe would break over the top of Sand Lake Hill, and let that Jimmy wind up. The road down the east side was not as crooked, but it had a couple of curves.

One day I had just topped the summit of the hill with a big load of logs, and rounded the first curve. There ahead was the small Jimmy with its load spilled all over the road. I got my load stopped. Paul, the driver of the Jimmy, stood there looking confused. He didn't see how it could have happened, he told me; he had been passing a car there and his trailer wheel caught the soft shoulder of the road. One of the logs had rolled into a cow pasture 50 feet below the road on his right, the rest were sprawled across the road.

I told him we better do something quick as Abe was next! He was loading when I pulled out. I knew Abe could never stop unless he was warned far up the road. I started back up the hill as fast as I could run. I could already hear that big, cackling Jimmy talking as it broke over the hill and Abe started feeding it gears as fast as he could. I could tell when Abe was satisfied with his speed and got off the throttle by the way the exhaust settled down to a roaring crackle as the weight of the load propelled the truck down the hill. I reached the curve in time to see Abe coming at me 50 miles an hour. I gave him the "hold it" signal with both arms. Then I jumped for the shoulder as Abe passed me like he was going to a fire, though I knew he had every ounce of brakes already set. I ran around the curve after him, and had a clear view of everything that happened. Abe ditched his truck on the left side, the only place he could go. The ditch was mucky and running with water. The Jimmy layed over, half on its side, and slogged along, rubbing the bank on its left, and ran 300 feet before it came to a stop, with Abe lying on the floor. When he ditched the truck, he knew the danger of the logs running forward and wiping out the cab. The slipping logs did come up snug against the cab but they did only minor damage. Paul opened the door of the truck and Abe crawled out. I ran back up the hill to flag the next truck. In 30 minutes we had most of the trucks lined up behind us, and traffic bottled up ahead of us. We pushed Paul's load off the road with log jacks, to be picked up later.

Using the jacks under the front of Abe's load, we raised it. One of the oncoming trucks hooked onto the

Yarder has pulled big log on landing close to loader (yarder not visible). *Courtesy Lila Boge, Tillamook, Oregon.*

Approaching log dump in Tillamook with a good load on a GMC.
Courtesy Joe Bell, Tillamook, Oregon.

front of Abe's truck and pulled it forward a few inches. Then several trucks, some pulling at an angle, and some forward, horsed the big Jimmy out of the ditch and onto the road, trailer and all. Abe hadn't even lost his load. He continued to the log dump and came back for more. His driving habits hadn't undergone any drastic change because of the incident. He was still "Wild Man Abe" driving his GMC like there was no tomorrow.

Some of us will remember the Sand Lake operation as another one of those *bad luck sides.* The biggest log I ever had on any truck, anywhere, I took out of the Sand Lake side. It measured 108 inches in diameter. Everything went fine until I reached the bridge that went steeply from the flat to the opposite high bank of the Tillamook River. The truck ran out of power near the top. I set the air, and had no trouble holding it there; the problem was how to get up the rest of the way. The dump was just around the bend in plain sight, and there were no trucks there. Had there been a truck available, a pull would have helped enough to get me over.

The next truck following me arrived and stopped at the foot of the bridge. The driver walked up; all drivers learn to help one another, each knowing *he* may need help next time. This driver had an idea. A small, short timber lay by the end of the bridge; he thought he could put it crosswise behind the trailer wheels as a back stop. Each time I gained a few inches, he would shove the timber forward so I could let the load drop back on it, and catch it on the rebound for another pull. He put the timber in place accordingly, and I revved up the engine, and let out the clutch. The front wheels nearly left the ground, but I gained two feet. As I kicked out the clutch, I could feel the wheels settle back against the timber. Catching it on the rebound, I threw the power to it again, and gained another three feet.

We repeated the procedure successfully over and over, until I reached the point on the hill where the truck could continue to move the load. I dumped the monster log into the river; then I went to Tillamook to have a new clutch put in. The only good fortune connected with that load was that the highway scales were closed that day. I had passed that scale so overloaded they would have put me in jail, and thrown away the key.

There were more spilled loads, more accidents, more trucks breaking down on Sand Lake Hill, and other unfortunate things while we worked that side than I could tell about. For example, it was the responsibility of the second loader to pull the rear reach pin and set it back a couple of holes. This was done so when the truck turned, the reach could run through its channel to follow the stinger action. A stinger is, in effect, an extension of the truck frame designed to lead the trailer almost exactly after the truck. Unless the reach is free to render through the trailer channels, the truck cannot turn.

I had a three-log bunk load one day with two on top. The loader forgot to pull the reach pin. The center log was exactly over the pin so when I did discover the trouble I couldn't readily correct it. Not encountering anything but gentle curves, I made it all the way from the landing to the junction of the Sand Lake road with Highway 101. That turn was sharp and uphill, and I couldn't turn. I maneuvered the truck with what little play I had, up and down the hill, getting straigtened out a little more each time in spite of the front wheels sliding instead of turning. But just then Al Coats, part owner of the Company and in charge of logging operations, drove up. Seeing my truck on the hill, with other trucks collecting behind me, and not having the faintest idea what the problem was, Al jumped to the solution most obvious to him. He ordered me to get the truck out of there, and off the job. The trucker next behind me asked what that was all about. I answered that I was fired. He wanted to know how come, and I said I was fired because the loader didn't pull the reach pin. He asked why I hadn't told Al, and I said he didn't give me a chance to tell him anything. He said *he* would tell him, and see that the loaders told him too.

He brought his log jack, and I got mine out. We pulled off the chains, knocked the cheese blocks loose, and pushed the load off into the ditch. I never did find out what was said, or by whom, I only know that the next morning I went right back out to Sand Lake and loaded up, and continued to do so. There was one other thing. Later on, after there had been a shutdown due to a strike, Al Coats came to my house to see if I would still haul for Coats, and I was the first trucker he came to.

Although typically somewhat taciturn, loggers are gregarious human beings. This was true of Roy, my nephew by marriage. He had worked on the rig-up crew, and in the rigging, almost the whole time I had been with the Coats Company. Roy had only been married to my niece a few months when we began at Sand Lake. He was a strong young man, well adapted to setting chokers. The work did not wear him down as it did some men, yet Roy had tired of the chokers. Several times he had asked the foreman to let him work at the landing. That was where people and trucks were coming and going. That was where the skills of loading were practiced and perfected, to the point that each load was made up of carefully chosen logs that would make desirable combinations to achieve a tight load of proper weight balanced perfectly on the truck.

One morning at Sand Lake a man came up the hill, avoiding the flying rigging, to the place where the rigging crew was working. The choker setters had just moved out of danger and given the signal that started the turn of logs to the landing. The new man asked which one was Roy. Roy spoke up and the man reported that the boss had said to come down and work at the landing. Though Roy had been after the boss for several weeks to let him work on the landing, strangely he hesitated, then said he didn't want to go down there, but he started down.

Since the day had just started, most of the trucks were still in the parking area. I was talking to another driver, and didn't see Roy come on the landing. After the truck that was being loaded, I would be next to load out.

The puncher of the loading donkey was having trouble getting a log to lie on the load just right. There were shouts of disapproval from the head loader, and we could see the hand signals, but we were paying little attention. Then the load partly separated and a log popped out toward the brow log. We heard a shout, excited voices, and the loaders picked up the log that slipped. Instead of putting it on the load, they signaled the puncher to swing it back on the pile. Then the loading donkey was shut down. These things attracted our attention at the trucks, because the loading donkey was a gas donkey and the motor normally ran all day except for the half-hour the puncher took for lunch.

We saw three men coming toward the parking area with the form of a fourth in their arms. I saw that it was Roy. He hadn't been on the landing fifteen minutes. He had been walking between the load and the brow log with a branding hammer in his hand when the log came out of the load, catching him between it and the brow log.

Roy was conscious as the men placed him, as gently as trained hospital staff could have done, into the back seat of a car. The car started for the Tillamook Hospital. I dropped my trailer and headed for the home of my sister, where I knew Roy's wife also was at the time. When I got to the house, I got out and approached the door. Hazel had already opened the door to meet me. I told her about the accident, and that Roy had been taken to the Tillamook Hospital. Hazel turned and ran out the walk to meet her daughter, several months pregnant, who was outside. I took them both to the hospital. Roy died shortly after they got there.

The loader puncher was so shaken by the tragedy that he told the foreman he wanted to quit; somehow he blamed himself.

I believe loggers tend to have a little more sympathy for others and a little less concern for themselves than most people. Who but a logger could make fun of himself for walking backward off the top of a load of planks, while gritting his teeth from the pain of a broken arm? Who but a logger, while being carried out of the woods so broken up that bones protruded from the flesh, would show concern for the four men trying to make their way through the brush and over logs to get him to where the ambulance could reach him, saying repeatedly, "put me down and take a rest boys" while the stretcher dripped with his life's blood. I'm told it's different now, but in those years every logger knew he might not be around next week, or even the next day. He was concerned about others as he knew others would be concerned about him.

While writing this, I returned to Sand Lake among other places, and was surprised at the changes. The road over the hill is much less steep, and wider. I had trouble finding where our logging road had joined the county road, so overgrown with brush was the area, but I did find it. I remembered so many things. I recalled how some of us had worked a 16-hour shift during the long summer days. Afraid of being shut down because of fire danger, the company had gone to two shifts. The first load was started at four in the morning, the last was finished at eight in the evening. To some of us, this had been a chance to get our bank accounts a little further in the black by driving both shifts instead of hiring another driver. After the last load was dumped, about nine in the evening, there was still the truck to be greased, checked over and fueled up for the next day. Drivers swallowed some supper and crawled into bed about ten-thirty. At three a.m. it was time to get up and head for the woods far to the south without bothering for breakfast. After a few weeks of this, men got tired, a chronic tiredness that affected health and tempers.

Drivers passed each other often with their doors partly open and leaning out so the wind would blow in their faces. They whistled, they sang, they did everything they could to stay awake. At the landing it became standard practice for one of the crew to walk over to the next truck in line and wake the driver. It was no time to aggravate a man either. A scaler found that out when he tried to cheat one of the truckers on his scale at the dump. The trucker threw him into the river. Some men who were demanding that we go out on strike found it out when they attempted to stop trucks on the road to force us to quit hauling. They found themselves facing an ordinarily mild-mannered man who picked up a double-bitted ax from his jockey box with the firm declaration that if they even got close to him he would use it on them.

Donkeys pounded, barked and whistled from daylight to dark. Men in the woods had riggin' in their hands as soon as they could see to pick it up. Logs came crashing into the landing just in time for the chaser to pull the chokers loose before two loaders were already jumping at them with tongs. A steady stream of logs went to the water 16 hours a day. Then, one day, all the feverish activity ceased. A strike had actually come. It was the millworkers who walked out, but logging stopped too. Coats had enough timber in the water to run their mill a long time, once the millworkers went back to work. No one could predict when they ever would.

Undercutting a ten-footer. The truckers who hauled the logs from such trees as this had no reason to complain. *Courtesy Lila Boge, Tillamook, Oregon.*

CHAPTER 15
GYPOS

The little outfits, those who logged out small stands of timber wherever they were to be found, were not closed by the strike. One of these, a man named Hall, knew Coats was down and called me.

He said he was rigged and ready to go again, and asked how would I like to haul his logs. I said sure I'd haul his logs, and asked him when he wanted me to start. He said tomorrow morning, if it was all right with me. The next morning I was out there. Hall logged and loaded with a single old steam donkey. He could consistently keep enough timber logged to the landing to keep me going to keep up. He had only one man working besides himself, and that was the donkey puncher. Hall was an extra good man to haul for and worth the hauling to get to watch his performance at the landing.

He would cry in wild supplication to the donkey puncher who was trying to maneuver a log into position on the load with the "crotchline" rig. The loader would keep trying and trying while Hall kept begging and pleading. After invoking the assistance of a variety of deities nobody ever heard of against the harrassed puncher, Hall would step back from the load and kneel with his supplicating hands held high while he pleaded passionately, as I recall, "pulleeaase! don't do that *again*! don't do it again!" Once the load was all in place and ready to chain, Hall would proceed about his business around the landing unperturbed, chatting amiably with the donkey puncher as though nothing unusual had happened. Considering the importunity of his pleas to the gods of hail, thunder, timber, and steam, I was always a little relieved that none of his requests were granted.

Hall soon had his little stand of timber logged and while he looked for more, I went up the Miami River to another gypo outfit I knew about. They put me to work all right, and I had already learned not to expect gypos to provide the same quality of roads that Coats had. Still, I didn't expect the up-and-down, tilted, crooked, slippery road they had. One man who had hauled for them for some time had been off the slippery planks several times. The problem was the road was not ditched along the side and in wet weather mud would wash down off the hill onto the road, a road that tilted toward the river. It is surprising that "grandma", as their regular trucker was called, had not lost his truck into the river. That old boy wasn't afraid of any road. The reason he was called "grandma" was that he was the fussiest man west of the Cascades about the way his truck was loaded. He had a hybrid outfit, a Chevrolet truck pulling a monster trailer that could carry twice as much as the truck. He wanted his loads balanced just that way, two-thirds on the trailer and one-third on the truck. Since the timber only went to the bay at Garibaldi, there were no highway scales to worry about.

There were times when there were no logs to haul, which is normal for a small operation. They have to move a donkey, or build more road, or splice lines, or dicker for more timber. On a haul that should have easily given a trucker six or seven loads a day, you would get two. Three if you were lucky. Then there would be days, and weeks, when you would get none. But I found a job to fill in. There was a little mill cutting alder lumber for furniture. It had to be hauled to Astoria. I would drop my trailer at home, remove the log bunk from the truck, and back under a big flatbed that I kept suspended with a winch. When there were no logs, I hauled lumber. When I was caught up on the lumber, maybe there would be logs again.

Then, one day I got a call from Forest Grove. A new outfit there wanted five trucks for a five-mile haul. I made no promises, but I rounded up a few of the unemployed truckers and told them about the haul. We decided to have a look at it. We went with all our outfits ready to haul. One trucker said that after we'd seen the haul we might want to turn around and go home, but maybe we could move a load or two to pay expenses. It did seem that if the haul was any good, there should have been plenty of trucks available around Forest Grove. Considering that, we still headed for the valley to see for ourselves. Stopping at Carlton, we slept in our trucks. It would be easier to find the road that had been described to us in the daylight. In the morning we found it.

Used to small operations with poor roads, we still were totally unprepared for the road we climbed to get to the logging. The road had been hewn out of the side of a hill by a 'dozer. The method was normal, but the result was not. There were turns in that road that shouldn't happen to a snake. But we had not seen the worst going up the hill. We had to go down a steep pitch to the landing. After a load was put on, the trucker had to pull up that pitch on plain dirt. It was completely impossible, of course. The gypo was told that. He proposed a plan he thought would work. Hanging a block on a stump at the top, he ran a line through it and back to the donkey. He would try pulling each truck until it had passed the worst of the hill and then stop pulling. This would let the hook drop from the truck and it would be on its own. We all had misgivings about the whole setup.

We got directions to the log dump anyway, and the first truck backed under the tree. The loaders put a light 2000-feet on that one. The hook was hung on the truck and with the help of the donkey it went over the hill. The second truck got a load of 2600 feet. It also made it up with the donkey's help. 3000 feet of timber was put on the next one. It barely made it because the donkey dropped it too soon. The fourth truck took a load of 4000 feet, awkwardly loaded heavy to the rear, and again the donkey dropped its pull too soon. The truck stalled and came down the hill backwards. The line was hooked on again, and the truck was horsed over the hump. My truck was last, and my load still bigger. I told the donkey puncher to keep pulling until he was sure I was over the steepest part. Afraid that he would pull too long and the line would then fail to drop, the puncher made his same mistake again. He slacked off too soon. I came down the hill backwards and before I got stopped I had almost rolled past the donkey on the wrong side and into a canyon. I told the boss to either get somebody up there to signal the donkey puncher when to slack off, or take the wood off and I'd go home.

The next time we had a go at it the puncher hung tough, and a man at the top signalled him when to drop me. Far from being the biggest truck there, my little rig staggered up the pitch and down the other side with a load far bigger than any of the others. All the others had had an awful time getting down the hill, but I didn't know that then. Fortunately the road was soft. It was easy to hold the load back, though the hill was steep. Creeping along in compound I negotiated the first couple of turns without trouble, but I knew the steepest part of the hill was still ahead. I dreaded the sharp turn just at the top of the steepest part. I was worried about Abe. He didn't have brakes enough for a hill that steep. But I remembered that his load was one of the lightest he had ever hauled, and thought he might have made it all right. I didn't know it then but all the trucks had lined up at the bottom of the hill waiting for everybody to get down. They had seen enough and only wanted to dump their loads and go home.

The turn I dreaded loomed ahead. Where the road had been carved out of the hillside, I saw truck wheel tracks clear up on the bank. The turn was just too sharp for a loaded truck. Later I learned that every truck had had to back up on the turn to make it. I crept into the turn with my steering wheel hard to the right. It wasn't enough. The left front wheel started to climb the bank. I stopped. I knew the truck had power enough to back the load up a few feet. Turning the front wheels to the left, I slipped into reverse, popped the power to it at the same instant I let out the clutch. The truck went back all right, but I heard a rending, splintering sound that made me a little sick. I set the air and got out to investigate. The heavy timber stinger had broken off where it protruded from the frame. I looked the situation over and decided that I could remove the stinger and run the reach ahead to hook it to the truck.

I got the stinger out of the way; then I began wrestling the reach, working the air hose back on the reach, and the reach forward through the trailer. I had it almost to the pin position when I heard a violent hissing sound. In spite of my efforts to prevent it, the air hose had come off the nipple at the trailer.

I knew the truck was going, and there wasn't a thing I could do about it. I scrambled out from under the load in time to avoid the wheels of the trailer as the truck and its load went down the hill.

I saw Abe coming up the hill on foot to see what was holding me up. Not knowing my truck had no air, and was not even connected to its trailer, he ran to try to get in the truck, he thought he could stop it. I screamed for him to get away from it and let it go. Abe looked puzzled, but he stayed clear. Well, at least I had kept Abe from getting killed. The truck left the road and turned on its side against a stump. The load tumbled off, and roared down into the canyon rolling over and over. I asked Abe to give me a cigarette. I hadn't smoked for five years.

At Forest Grove I ordered a heavy wrecker to bring my truck in. I contacted the insurance company and rode one of the other trucks back to Tillamook. In a couple of days I went to talk to Everett Condit who owned the Chevrolet Garage in Tillamook. I told him I'd heard from the insurance company, and he asked what they had to say. I told him they were awarding the rebuild job to the garage in Forest Grove. Condit asked if that was the garage that towed it in and I said it was. He told me that I didn't have to take it, and I wanted to know how I could get out of it. Condit said that no matter what the insurance company told me, they couldn't make me have the truck fixed any place where I didn't want it fixed. I asked if he was sure of that and he said he supposed they could hold out for the *amount* of the bid, but they sure couldn't make me have the work done in Forest Grove. I wondered how we could get the truck back to Tillamook, as Condit had no wrecker. He pointed out back where a truck sat with no body on the frame, and asked if I thought I could get it back with that. I said sure, that I would take a man along to help me load it.

The next morning Doc Hatton went with me to Forest Grove. Entering the garage there, I asked about my truck. The shop foreman said they hadn't started on it yet, but that they had the job. I told them I had come to take it to Tillamook. The foreman sent for the manager, and they went into a huddle. The manager came over to us and informed me that I couldn't take that truck out of there, that their repair bid was accepted by the insurance company. I asked if I had to get a court order and he reiterated that the job had been awarded to them. I told him it was my truck and I'd have it fixed where I wanted, not where the insurance company said. He growled that there was a tow bill against it. I said that I'd write him a check, that the insurance company would have to pay in any case.

A few minutes later Doc and I had the old girl reared up with her front end sitting on the naked frame of the old 33. We chained her down so there would be freedom to turn but it couldn't go forward or backward or sidewise. Then we pulled out for Tillamook.

There's not much more to that story. In three weeks I had a truck just like new, and the insurance paid the bill. I was ready to get back to hauling logs again. I had learned from this experience. Bad as it is for a truck to sit idle, there are some jobs a man is a lot better off without.

CHAPTER 16
CRUSHED ROCK AND LOST LOGS

I learned that a new logging outfit was going to start on the new Sunset Highway just being built. I went to Seaside and inquired further. I had heard that several hauling bids had already been made for the job and because work was still scarce all over the coast, the bidding was running very low. I headed out to where the logging show was on the upper Nehalem River. Stopping near the end of a low bridge, I walked across to where a man was standing with a shovel in his hand. I said I was looking for Mr. Corey, and he replied that I was looking at him. I said I had heard he had some timber to move, and he mentioned seventy-five million feet. I asked about the road, that the coarse ballast rock would cut tires to pieces. He said the County had promised to put topping on it. I told him I understood he had had some bids on the hauling, and he answered yes, and asked if I wanted to bid also. I said it might depend on whether he was going to take the lowest bid or not, because I had heard that some of the bids were awfully low. Corey asked how many trucks I had, and I asked how many did he need. He said ten, that if I could provide that many, we'd talk; if not there wasn't much use of talking. I told him I didn't have them, but could get them. He then asked if I intended to subcontract, and I said that was right, that I'd put two of my own on the job and get eight more. Then he asked what was my price for the ten-mile haul from the woods to the Kerry Line. I said two and a quarter. He exploded, said he could get them hauled for a dollar and a half. I told him a contractor would go broke at that figure. Corey guessed that would be his problem. I pointed out that it would be Corey's problem if the contractor was not able to afford normal upkeep, tires, and so forth, that he would find himself with timber to be hauled and trucks broken down. Then he'd have to get a new contractor.

We talked for an hour, including the time spent looking at the timber. I saw things I liked and things I didn't like. At the landing there were two fine steam donkeys, probably left over from Corey's railroad logging days. The road had certainly been built by someone who didn't know much about building truck roads. The timber had been burned, a casualty of the great Tillamook fire, but I knew that if we got it out in a year or two, worms would not have damaged it too much.

Though I was fifth from the lowest bidder, I got the contract. Once again I contacted other truckers who were not working. We put together the ten trucks and moved up near the job. I would have laughed at Corey's estimated need for ten trucks, if it had been anybody else. But I knew his reputation from years back. I knew that his outfit would keep a steady stream of logs pouring into the landing from the starting whistle to quitting time.

I had to manage to get another truck and trailer of my own, and two drivers. I had already put a dual axle trailer behind the other one, now I got a new Chevrolet and a dual axle Peerless trailer. I bought the new one equipped with an Eaton Ruxtell two-speed differential and had the same thing put in the truck I already had. That way we could load with the butts back, and two-thirds of the load on the trailer.

In Tillamook, Stone's logging works had already gone into the burned timber up the Wilson River with a string of Fords equipped with dual axle Grayco rear ends, and pulling dual axle trailers. But that was about the only such fleet in Western Oregon, and Stone was having bad luck with light Fords pulling so much wood over such a soft road with the dual rear ends. All the rest of us, seeing the trouble Stone had, decided to stick with the single axle trucks. Our roads were soft too.

When we crossed the river from the logging road to the county road, we had to cross an underwater bridge, the only one like it I had ever seen. The bridge was built so the water would go over it when the river was high, and under it when the river was down. There was a short, steep pitch from the floor of the bridge to the county road. There was only one way a truck could get up that pitch with a load of logs, other than being pulled up by a cat, which was done in a few instances, before all the truckers got used to the spot and learned to negotiate it. I had taken a long look at it before I came out with my first load.

Coming down the road with that first load of nice fir timber, I kept my speed up and rounded the curve onto the bridge as fast as I dared, really throwing the power to the truck, running at the opposite bank in third gear as fast as possible. When the truck hit the sharp incline it quickly lost speed. I skipped a gear and went directly into compound and layed the power to her. The truck barely dragged the load over the top. If a driver tried to drop his gears one at a time, as he would ordinarily do, the two seconds of loss of power would be too much, he would lose it on the hill and go down backward, either onto the bridge, or into the river. I know it, because it happened. We started hauling in wet spring weather which made it even worse. The roads were soft, partly because they were new, but mostly because they were water-soaked. In the course of delivering a load of logs to the dump, a truck was *never* in high gear. Returning it might be a small part of the time, because the rig then was not heavy enough to *bend* the road, and had only the muddy surface to contend with. I have stood beside that new road and watched a heavily-loaded truck go by, and watched the road sag under the weight of the drivers, and even more under the trailer where two-thirds of the weight was.

But that road, the section of the new Sunset Highway, was not our worst problem. The county road between the logging road and the Sunset Highway never was topped with fine rock by the county, as promised. There we were hauling over a road with only a heavy layer of ballast rock on top. The pieces of crushed rock would run two inches or more in size, and had sharp edges that cut tires to ribbons. The first month we hauled over that stretch all ten trucks had to replace all the tires except a few on the front wheels.

It turned out that we dumped for reload on the railroad at the very site where the B-W camp had been a few years before. I remembered the days when I worked in the rigging there, and my brother-in-law ran the big Baldwin. I had been warned by several friends not to go on that job. All the stories about dead men being stashed behind stumps in case there might be others were revived. I even heard some new ones. I discounted them as exaggerations. One thing I could believe was that the crew sure knew how to log, and knew how to put a good load on a truck in a hurry, which helped. But good outfit or bad, accidents will happen.

One day my best friend, from Salem, showed up. He was a timber faller, possibly the best in the state. He came to my house and talked. He had heard the same "highball logging" stories that I had, but he figured that as a faller his safety would be up to him. I agreed, and Les asked for a falling job the next morning. He got it. The second day on the job he was being carried out; a tree he had dropped had brushed a snag, the top popped out of the snag, and though Les was watching the top as it kicked back toward him, somehow he didn't manage to get out of the way and it broke him all up.

Returning from the dump I heard the news. I headed for the hospital in Astoria. I was not allowed to see Les that night. He was dead before morning.

One of my drivers, whom I had told several times to be sure he was in the clear when he cut a load loose at the dump, failed to do that. A log kicked back off the truck and hit his lower leg just as he tried to jump clear. The log drove his foot and ankle right into the gravel road. He lingered in the hospital, and finally was able to go home, but he didn't live long. From then on, my brother drove one of my trucks, and I drove the other as long as we logged there.

One of the drivers I had brought up from Tillamook liked to load light. He didn't believe in overloading like the rest of us. He may have been right, his truck certainly lasted a long time. He was at the tree one day when we other drivers watched jealously while the loaders picked up a fine big log,ideal for a one-log load. As the big log swung out over Smith's truck, the man actually turned pale, and said in a hoarse voice, "my God, boys, don't put that log on my truck".

We logged like it was going out of style, pulling the guts out of our trucks on hills and stopping for a tire change every other trip because one of the sharp rocks had cut its way clear through a tire. But we weren't making money. The haul was just too tough, in too many ways. Smith, with his light loads, fewer repairs and less tire damage, probably made as much, net, as any of us. I don't think we could have made much at three dollars a thousand, let alone two and a quarter. One of my subcontractors was so behind in his payments and his truck was so rundown he begged me to drive it one trip and at a certain place on the cliff above the river just step out and let her go. I told him if he wanted to swindle the insurance company he would have to do it himself.

Did we think we had it tough? We hadn't seen anything yet. First, when we got our log tally back, on the basis of which we would be paid, it was always

Author's two trucks and their drivers at the residence at Elsie, Oregon on the Sunset Highway.

numerous logs short of the number we had hauled, and we had *kept track* of. They were listed as lost logs. Where did they get lost? Did they all fall off the cars as the train took them to the Columbia? Did they get out of the boom and float away? Most of us didn't buy any of these possibilities. We thought we were plain being swindled.

Then the sky caved in. The thing that had been threatening for two months happened. the well-remembered Columbia Basin strike occurred. Everything came to a grinding halt. Most of the truckers had truck payments to make and insurance payments to keep up, to say nothing of living costs and other unavoidable expenses. With thousands of trucks off their jobs, there wasn't much chance of picking up incidental hauling. We were in a situation not of our own making, over which we had no control. We lived with the fading, glimmering hope of finding something to do with our trucks. It developed that the strike did not affect all companies the same. *Some* of those whose logs did not have to be moved over railroads were still working. One company with a mill in Portland was logging on the west end of the Sunset Highway, hauling from the summit of the coast range of mountains into Young's Bay near Astoria. Somehow, in spite of all the other idle trucks, I managed to get my two trucks on that 27-mile haul.

We were not bothered much with the soft road on that job, because we had no adverse grade on any of it going out loaded. From the junction of the Sunset Highway with 101 the road was all hard-surfaced. Although it was a long haul for those days, it was not a hard haul. We could get four trips by dumping the last load late. The highway crossed the railroad at a point between Seaside and Warrenton. There were no semaphores, and view of the crossing was cut off approaching the railroad from the west. The approach to the crossing was also downgrade. None of those things excused what happened to one of the truckers on our haul.

He approached the Southern Pacific crossing with a long heavy load of logs. The noise of the truck effectively smothered the sound of the train whistle. Chuck, the driver, was running down the hill with full power to help him over the little hill on the other side of the tracks. Suddenly the engine of a Southern Pacific train appeared at the crossing. The train was almost in front of Chuck before he saw it. He knew the futility of trying to stop. He opened the door and bailed out. Twisting and rolling, and tumbling, Chuck's body stopped on the ends of the ties parallel to the track, with the cars hurtling by almost on top of him. His truck rammed the train just behind the tender. Pieces of truck and chunks of splintered, broken logs were dragged and scattered down the track for 200 yards south of the crossing. Chuck had no serious injuries, though he was scuffed and bruised all over his body. He not only never hauled another log, he never drove anything again.

It soon became evident that there was no money to be made on that haul. Really, we had known it when we started, but like a lot of other truckers, we were desparate for something to do. The first problem was that our pay was extremely slow in coming. The worst problem was that when it did finally come through, we found it to be a repetition of our earlier problem; *many* of the logs we hauled were listed as *lost* . Finally the company had all the logs they could handle, they had no booming grounds for more. The shutdown came as *no* bitter loss to us because all we had been doing was wearing out trucks. If we could have been paid for *every log* we hauled, we might have made all expenses plus a small margin.

The road back down from success was a bitter one. Property accumulated over several years was mortgaged, sold or traded to try to keep afloat. Everything possible was done to try to keep truck payments up to within sixty days in arrears, to avert repossession. All the truckers were in the same boat. One by one they went under. The Columbia basin strike was virtually solid by then, affecting non-striking companies because there was no outlet for their logs. With tens of thousands of loggers and thousands of trucks unemployed, the *strike* lasted three months. After the men were starved out, willing and eager to go back to work, the timbermen and mill operators refused to go back into operation, so a *lockout* was in effect for more than a month more. They wanted to teach the strikers a lesson.

Like a gambler, looking at a bad poker hand, I was looking back at a six-week winter shutdown, four months of unprofitable hauling, the accrual of additional expense of one more truck and trailer, and finally the almost total lack of income. It was a slow death. The day came when I realized that I could not possibly hold out. The dealer from whom I had bought the trucks told me he had to have money, and lots of it. I told him I had borrowed, had sold property, and had mortgaged my home for every dime I could get on it. I told him he knew very well that as soon as we could get back to work I'd be all right. There was more talk, ending in this final proposal: the dealer said to give him back one of the rigs, so he could sell it and get the bank off his back, and he'd hang tough on the other one until the strike was over. I wanted to know which rig and he said the new one. I told him to take the old one and I'd try to get back on top with the new one, but he said he couldn't do that. The new one was the only one he could sell for enough to get the bank off him. I saw there was no changing his mind, and I didn't even blame him. God knows he had his problems too, I was only one of many deliquent accounts. Then I said I would turn them both back, period. I wouldn't try to pull out with a truck that was soon going to begin needing repairs.

The following morning my brother and I headed for Tillamook with both trucks. I almost hated it more for my brother than I did for myself. He had been driving the new truck. He too had had a hard time through the depression and I had hoped that we had something going that might mean eventual independence for both of us.

After we returned from Tillamook, we happened to be in Astoria on business. Stopping by the International agency, I was surprised to be offered two new outfits, medium-sized Internationals with two-speed rear-ends and my choice of dual axle trailers. I was told I could take delivery just before I was ready to go

Load like this can only be legally hauled on private road. (Trucker unknown, used only as an example of heavy loading on private road.)

back on my haul, I wouldn't have to make a down payment, and I'd pay 25% of my gross until the trucks were paid for. I said I'd give it some thought and let him know.

I did give it some thought. The thing I felt most at that time was that I had had it with trucking. I didn't believe that it was any longer possible to make money with trucks. Bids on jobs had become "cut-throat". State controls on the minimum amount that could be paid for hauling had come to other states, but not yet to Oregon. There seemed to be no way to keep operators honest, your profit and even wages could go down the tubes under the heading of "lost logs". Mostly I was just unreasonably discouraged. Pulling together what assets I could, including an equity in a Tillamook apartment house, I sold everything and headed for Florida. I told myself I would never log again.

A few months after I had left Oregon I learned that one of the truckers who had hauled under my contract had contracted with a logger on a private haul (no public roads). He had managed to get hold of several Mack trucks. He had no highway scales to worry about, the timber was good, and the road was good. He would be financially independent in a few years. Why couldn't I have found a setup like that? I knew logging, and I knew trucking. I knew it was true that I was as capable as anybody. With the right breaks I would have made a success of truck logging, where some others would have failed. In the past I had done just that, but... it hadn't happened for me that time, and I think it was partly because I had violated the most important part of my own philosophy, "*never* give up". Anyway, I kept telling myself, *never again.*

Timber like this, with no load limit, can make a man money.
Courtesy University of Washington Library, Kinsey Collection.

CHAPTER 17
LOG FEVER

It was years later that the turns of destiny found me in the heart of timber country again. I was running a shop in Forks, Washington. So much water had gone under the bridge. There had been a couple years of working at whatever I could find. There had been a couple more spent perfecting a trade I already knew something about. There had been a couple of years helping fight a war, though I was in Africa, not Korea.

I liked the wild country, and my body and paint shop was doing well. Whenever I had a day off I was out in the timber with my son, Dan. We hunted bear, and birds, and saw thousands of deer and many herds of elk. We made friends. Yet I was not completely satisfied. Most of my friends were loggers. I couldn't help going out to some of the logging shows once in awhile, while they were working. There was something about the sights and sounds of logging that drew me in spite of myself. I liked to watch the loaders build a beautiful load on a truck. I looked at the trucks being used. There wasn't a Chevrolet or a Ford in the bunch. Weight limits on the highways were much higher. I saw my first Aberdeen stakes, they stood four feet high replacing the old eight-inch cheese blocks. With them a load could be built higher, and it was safer. The load limit could easily be reached. I admired the trucks with Cummins Diesel motors, Jimmy diesels and Budas. Transmissions had come a long way too. All trucks had dual ratio rear ends. All trucks wore ten-inch rubber. When my logger friends dropped by the body shop, we didn't talk about my work, we talked logging, and trucking. Every night I saw strings of trucks, already loaded, sitting along the streets of Forks ready to head for Port Angeles in the morning. Hauls were so long that all the trailers were loaded on the trucks at the dump for the return to the woods, where the loading donkey lifted them off. Hauling, instead of pulling, the trailers was cheaper and gave the trucks additional traction up the hills into the woods. Some of the trucks hauled into Port Angeles 50 miles away across the flat and around the shores of beautiful Lake Crescent. Others went to Hoquiam or Aberdeen more than 75 miles from their landings.

The fever of logging and trucking was coming back stronger every day. How I was tempted to take a shot at it once again. Then came the straw that broke the camel's back. One of my friends came into the shop. I painted signs on truck doors, but he didn't want that. I repaired radiators, but his didn't need repair. I straightened and painted bodies and fenders, but he didn't have anything bent. But he did have *something* on his mind. He knew where I could get a cedar haul, and I asked what made him think I wanted a cedar haul. He said he knew I'd never be satisfied until I was back hauling logs, so I asked him why he didn't take it. He said he wouldn't quit the job he had, which was hauling for the best outfit in the country. I couldn't deny that. I asked about some of his other friends, if one of them wouldn't like to have it. He said, sure, but it would give me a chance to get started. I pointed out the obvious fact that I had no truck, and he told me to quit fighting it, that he knew where I could get a truck.

We went up and looked at the cedar. It had been left at a time when the fir had been logged and there was no market for cedar. Now, cedar was selling high, and there was a lot of it there. I talked to the operators, and they told me that if I could have a truck ready to haul by the middle of the following week, I had the job.

Once again I plunged in and bought a truck with no down payment. I had a big White, and a fine trailer ready to go on time. It was equipped with Aberdeen stakes and Timkin two-speed rear-ends. These big trucks were quite different from the ones I had used. Empty, they would go down the road at 75 miles an hour, though the speed limit was less. Most of the power plants were Cummins. They far outnumbered the Budas and the Jimmies. The most popular truck

Author's outfit preparing to load a donkey. White truck, with heavy trailer, near Forks, Washington.

Completing the loading, using donkey's power. Author's White, near Forks, Washington.

Gas donkey loaded ready to go. White truck starting down the hill, beginning the twenty-mile haul to the new logging "show".

Preparing to unload donkey from White using blocks and the donkey's power to skid the donkey sidewise off the truck.

Unloading donkey from White, near Forks, Washington. Ten minutes later the donkey was in place and the author was on his way back to Forks, Washington.

was the Kenworth. The brakes were straight air, each truck having its own compressor. Vacuum brakes were no longer used.

At that time, 1952, there was only one steam donkey still in use in that part of the peninsula country, and it was fired by oil. The track loader was generally used for loading, though there were quite a few gas and diesel donkeys still in use, with heel booms, hayrack booms, and crotchlines. On the more gentle hills toward the ocean much of the logging was done with cats. In the "big hill country" the ground was too steep for them, except for use in building roads. Up there, they used big diesel donkeys and the name *Skagit* on the name plate was a familiar one.

Within a few days I had put a lot of cedar into the bay at Port Angeles, and into the pond at the Pleasant Valley mill. Then, without warning, the partners for whom I hauled stopped logging. One of the partners was around town drunk for several days, and the other one got a job in a garage. I heard several stories, one was that the partners had lost their stumpage. Another was that their creditors had attached their money at the purchasing mills. The money from my hauling came directly from the mills, so I was paid with no trouble. I never found out really why they had quit. What I did know for sure was I had a truck to pay for and was out of a job. I picked up a few loads here and there at random, then I heard of an Indian foreman, running a show in the big hill country who could use an extra truck. I contacted him, and began hauling for him the very next day.

That Indian was a magnificent logger. He was also a little reckless. He kept a D-8 Cat at the landing to enlarge the landing and keep it clear of the debris that kept accumulating as logs were dragged in by the yarder. I used to watch in amazement, while waiting my turn, as that man pushed a bladeful of dirt and rubbish over the bank, not stopping the cat until it hung teetering on the very edge. A few months later the cat teetered too much one day and cat and driver

Reloading truck to car, using hay-rack boom. Slings with rails at bottom enabled the movement of the entire load at once. *Courtesy Weyerhaeuser Timber Company, Hoquiam, Washington.*

Modern mechanized logging in Pacific Northwest. Loading logs from right of way on Lake Pleasant area of Clallam County. Heel-boom loader requiring single pair of tongs. *Courtesy Rayonier Company, Hoquiam, Washington.*

One of high-country logging roads can be seen near the snow line above this up-land valley.

went over the bank tumbling over logs and stumps to a stop far below. The world had lost a fine human being, and his company had lost a top logger.

The road into that show was good, but it was necessarily long, steep and crooked. The first problem I had on that job was that the front brakes on my White had never been disconnected. Brakes on the front wheels can lock and slide, depriving you of the ability to steer your truck when you need it most. Fortunately, it happened just before I started down the main hill. The left front wheel was forced back by the brake until it was against the back of the fender. I was able to get the truck to the nearest turnout so the other trucks could operate, then I caught a ride to town and got a new center spring bolt. After I had installed the bolt, I disconnected the front brakes so the same thing couldn't happen again.

It was something of a thrill, taking the first 8000-foot load down that hill. Used to plank roads, where you'd better not meet a truck except at the planked turnouts, I just about flipped when I saw the first truck come blasting at me around a blind corner, but the driver moved over until his outside tires almost hung over the bank, and I had room to pass him. From that road one could look down at one point where a truck still lay where it had finished rolling after its driver had "lost it" on the hill and jumped. No attempt had ever been made to salvage the truck, because of the nearly vertical nature of the mountain.

I soon got used to it. There were plenty of places where trucks could pass and on blind corners room had been made for passing on each side of the corner. My oldest son, Joe, happened to be in Forks on a visit once. He wanted to ride with me through a day's logging. When we met a truck right in the middle of a big cut with a blind corner at the end, I threw my truck right up onto the bank of the cut. He has never forgotten that, but it was the same kind of driving we all did every day.

Then, there was the "re-logging" every night either at Bear Creek, where the truckers would stop for coffee on their way home from the last trip into Port

Kenworth truck and Kenworth fork loader near Forks, Washington. This great machine could take an entire load from a truck and transfer it to a railroad car. *Courtesy Rayonier Corporation, Hoquiam, Washington.*

Angeles, or at one of the cafes in Forks. Loggers and truckers would gather to recount the events of the day. These "re-logging" stories were apt to be well seasoned with fiction. Some of them were *pure* fiction. But some stories were too strange to be fiction. There were a dozen stories about a man named George Raymond. George had a wild temper, as I've found many loggers have. If George signalled for slack in a line and the slack didn't come as readily as he thought it should, he would scream very profane things, and proceed to gnaw on the line with his teeth while he yanked and jerked at it until he lost his balance and fell down. George had the unaccountable ability to snap out of one of his rages as fast as he went into it. I saw this myself one day when I went to his big shop for some needed hardware. Outside by a small service building was George, tearing the shingles off with his bare hands from which the blood was flying. He didn't like the job the shingler had done. Thinking I had come at the worst possible time, I was prepared to leave, but George saw me. The twist of rage went out of his features. He turned to me, and as calmly as if he had just awakened from a nap, asked what he could do for me. I told him I needed some long bolts, but that maybe I had better come back later. He said no, to come on inside and he'd get me what I wanted.

Then there was the time George got into a fight with one of his men in the woods. Not understanding George's lapses, the man had taken exception to some of the things George was screaming, and had busted him in the mouth. In the ensuing fight, George came out second best, probably because of his blind rage. That night, in a Forks card room, George was approached by the man who had nearly kicked a lung out of him. Interrupting his card game, George asked genially what he could do for the man. The logger said he wanted to get his time. George told him sure, but he couldn't imagine why. The man said he supposed he was fired, and George said he didn't know what he would be fired for. The man seem greatly surprised. Typical of loggers around Forks, the man was back on the job the next morning. I guess he thought if George had no complaint, there was no reason *he* should have.

Another man, named Orval, drove one of George's trucks, hauling timber into Port Angeles. Coming home one afternoon, Orval knew he couldn't make another trip so he stopped in a tavern and got somewhat sloshed. Following the crooked contour of the road around Crescent Lake, he got sleepy. The truck left the road on a curve and plunged into the lake. By some quirk of fate, Orval was thrown out of the truck and landed spraddled out on a big flat rock, looking like a sleeping frog. The truck quietly settled to the bottom of the lake where it was so deep that recovery of it was highly unlikely. Long afterward one of us asked Orval about the incident. He said it was strange, but George didn't speak to him for a long time, that he wasn't mad or anything, he just couldn't think of anything to say.

In spite of his foibles, George was a good logging operator, and good to work for. If I ever needed a load while my regular outfit was down temporarily, I just drove in to Raymond's landing and lined up behind the other trucks. When they were all loaded, I would

Loaded truck on one of better roads near Forks, Washington. *Courtesy Rayonier Corporation, Hoquiam, Washington.*

These four "cornbinders" carry logs scaling up to 9000 board feet, not counting cradle logs which will contain hundreds of board feet. *Courtesy University of Washington, Kinsey Collection.*

pull under the tree and the loaders would put on a load. George saw me, he knew I was there without being asked. He never did say a word. That was the gentle side of his nature that balanced out his volcanic temper.

Orval, on the other hand, was something of a character himself. Long after the episode of losing the truck in the lake, Orval was surprised when George Raymond called him, wanting to know if Orval could run a Skagit shovel loader. Orval said he supposed so, he had run about everything else. George told him to come on down, then, that he had some logs back of his shop that he wanted to load out. Orval went down there and climbed up on the loader. He was six feet tall, broad and heavy. He started the rig up and swung it around to pick up the first log. The tongs were set and Orval swung the log over the truck. He was unfamiliar with the loader and was feeling his way. After a couple passes at loading the log, George started to climb up into the loader to straighten Orval out. Orval told him not to come up there but just stay out of his way. He got the log on the truck, swung the loader around to get another one, and again he had some trouble manipulating the levers so the loader would do what he wanted. Again George started to climb into the cab of the loader. Orval roared that he had told him once to stay down out of there. He continued the loading, but it wasn't long before once again George ran wildly to the side of the loader and grabbed the steel sill of the cab to climb up over the track. Orval picked up a big spanner wrench and cracked George solidly first on one set of knuckles, and then on the other. Once again he warned George about the folly of trying to climb into a loader that he, Orval, was running. George swayed and stomped and put his battered hands under his arms. He took them out one at a time, and blew on them. He didn't try to climb up any more, but for the next half-hour he was seen blowing on his hands every little while, and shaking them in agony. It never would have crossed his mind to fire a man because of personal pain.

One night Orval was in an unsocial mood. He had been having an argument with his wife. He was sitting on a stool at the local bar sulking. His mood was not being improved any by the loud voice coming from two stools to his left. The man kept making idiotic statements. It was clear he didn't have the slightest idea what he was talking about.

After awhile Orval's patience wore thin. He asked the man why he didn't shut up. The man on Orval's immediate left demanded to know if Orval was talking to him. Orval said sarcastically that he wouldn't waste words on any little squirt, and he back-handed the man, who was not at all that small, halfway across the room. He had to get him out of the way anyhow, because the loud-mouth on the second stool was going to be next.

Just to live among men like loggers is an unforgettable experience. They were crude and sometimes uncouth, but they were genuine. Sure, there were a few rats around, but they didn't fare too well with the others. When the state police scale was operating, truck drivers returning from Port Angeles would warn in-bound trucks by extending an arm, palm upward and making a lifting motion to indicate weighing. The in-bound truck, if too heavy, could pull off the main road and wait until the scale closed. This courtesy was extended to all drivers, no matter who they hauled for, with the exception of certain hated "rats". Other drivers wouldn't lift a finger to help them.

Truckers would sometimes personally react to the troubles of another. The Park Ranger at Crescent Lake got in bad with a couple of truckers by attempting to hold them to ridiculously low speeds through the park. He had no way of accurately establishing speeds, he only had his car which did not have a calibrated speedometer, but he would try to make it stick anyway. The ranger talked with a soprano lisp. Stopping one of the drivers one day, he shook his finger at him, like a mother might do to a naughty boy, while he piped tremulously that the man had been "thpeeding". The result was he had over a hundred truckers bent on heckling him to death. They nicknamed him Smokey, for the bear I suppose, and ran him ragged. Then one night the drivers' disgust was driven to a new level of nausea. Several of us were returning late from Port Angeles, and rounding one of the turns in the road I saw what I thought was the moon on the lake. I looked at the water again, and realized that the light was coming from the bottom of the lake. Then I saw a couple of trucks pulled off by the guardrail. I stopped too, and one of the drivers was standing shivering by his truck. He had recognized the light as coming from a submerged car. He had tried to reach it by diving; an extremely strong swimmer he still had found it impossible to reach the car. At the depth of thirty or forty feet the diver had hit a layer of water so cold as to almost paralyze his muscles. He had to give it up to save himself.

It was at this point that "Smokey" came into the picture. The Crescent Lake Park, which was under Smokey's jurisdiction, was equipped with a rescue boat and other equipment for such occasions. The boat had a powerful spotlight, and grappling hooks. But Smokey thought it best to wait until morning, so the newsmen could be present with their cameras. Everyone knew the man under the lake had to be dead already, but that didn't make Smokey any more popular. I remembered having seen the victim in Forks earlier, he was drinking then and the bartender had asked him if he had to drive around Lake Crescent that day.

Log truck drivers knew they had to stay within the law, particularly concerning weight, or they would pay heavily. It seemed to them that penalties for overloading were excessive; if several infractions had occurred, a fine could be as much as $500. It was even possible for a driver to have his permit suspended. Companies that owned their own trucks took the matter seriously. They didn't want fines or suspensions. Many of them put in scales near the highway with a loading rig there so a driver could have a log or two removed if his rig was too heavy, or some added if he weighed light. Drivers of these trucks could haul their loads to the dump without worrying about the State scales.

The smaller logging companies could not afford the expense of a re-loading operation and scale. For one thing they moved around too much. A scale installed in one place might have to be moved in a month or so to a new location. Drivers who hauled for those companies travelled with fear of the scale gnawing at them every other load. Some of those drivers became downright unfriendly toward State Police, especially if one or two of the drivers appeared to be victims of a vendetta. Some of the State Police used to carry "jump scales" in their

View of heel-boom lifting trailer from truck for loading. *Courtesy John Coats, Portland, Oregon.*

trucks. An officer would stop a truck and make the driver pull one wheel at a time onto the puny scale. With one wheel forced to lift four inches above the others, heavy springs had to flex and the scale would read hundreds of pounds above actual weight. One driver was stopped near Aberdeen and ordered to pull one of his drivers onto the jump scale. So aggravated and enraged was he because he felt his persecution had gone too far that he walked around the truck, picked up the jump scale and threw it out into the swamp. That didn't get him out of trouble, of course. He was charged for the scale.

One officer in the Port Angeles-Sequim area had developed a nasty habit of trailing loaded trucks after dark with his lights out. He followed close hoping not to be visible to the truck driver on curves. If the speed limit was exceeded, the officer would turn on his lights, whip out from behind his victim and pull him over. His bad habit was broken suddenly one night when he was tailgating a truck he had nailed once before. The driver caught a glimpse of the unlighted car in the headlights of a passing car. He didn't do anything until he came to a little bridge just as a car was approaching. Knowing the officer had no place he could go, the truck driver slapped on his air brakes. The stopping truck took the officer so much by surprise he impaled his radiator on the reach of the log truck before he could get on his brakes. The truck driver poured on the power and drove away. The officer was left in the road with a car that wheezed and emitted sizzling steam, and gurgling water.

Officers who were willing to meet the truckers halfway were respected and liked by all the drivers. The one who was permanently stationed at Forks was liked by all the truckers. When he saw a driver speed up to get a run at a little hill, he understood that the man had no intention of maintaining that speed, but was only trying to get over the hump without having to gear down too far, which would cost time, and time meant money. From Port Angeles to Aberdeen, every State Police officer was known to the truckers. They knew each man's habits of hiding, if any, and whether he was a reasonable man or a "ticket-happy cop". The log hauler was caught between a rock and a hard place. He lived by moving loads as heavy as possible, as fast as possible. The police officer's job was to see that he hauled neither too heavy nor too fast. Any driver who was trying to show a margin of profit was bound to run a little above speed once in awhile, especially while running back empty. He did not dare load too light consistently, and if he tried to load about to limit each time he was bound to be a little over on occasion. Looking at a long period of time, such as from the time a trucker goes in debt for his outfit until the time it has to be replaced, the average trucker operates on such a narrow margin that he is lucky to come out with decent wages, let alone profit.

I have known truckers who were only grown-up kids, with a monster toy they used to make a living. The men who had good trucks were inordinately proud of them. They were not above a little contest now and then. Two of the truckers were returning to Forks one night. They had run around the Lake as fast as they dared, and come to a long straight hill at the west end. There was no adverse traffic on the hill for it was late at night. One of the trucks pulled to the left and slowly but surely passed the other. They met at a cafe in Forks. The driver who had been passed accused the other of giving his "hog" a shot of ether on the hill. Some of the truckers *had* a small ether tank for starting on cold mornings. The other indignantly denied the use of ether saying his truck would beat the first man's "hog" without resorting to ether. When asked about the night before last when on the flat the other side of Bear Creek, he was doing a flat "six bits", the fast Jimmy came boiling by like he was backing up, and he didn't suppose the Jimmy driver expected him to believe that he had not used ether there either. He warned the other driver that he would "blow that thing up" doing that, and cautioned that ether was like nitroglycerine, in that a little bit goes a long way. The Jimmy driver doggedly insisted that he never used ether once the truck was running, but that his rig would just outrun the others, and that was all there was to it. So it would go; each driver pretending great indignation, while they sat drinking coffee together. Kids with oversized toys, enjoying every minute of it.

Loading under a heel-boom using single pair of tongs. *Courtesy John Coats, Portland, Oregon.*

Mixed west coast timber. *Courtesy Oregon Historical Society, Portland, Oregon.*

CHAPTER 18
WILD COUNTRY

In 1952, in spite of the inroads of logging, the Olympic Peninsula was still a wild, primitive country. So much of it is contained in the Olympic National Park, where hunting is prohibited, that the areas surrounding the Park are also populated heavily by game and non-game animals. I once read of a bus being halted by a herd of elk crossing the road, but I didn't think I would ever have the same experience. However, one night two of us were returning from Aberdeen late. My partner, who was driving, poked me and said excitedly that I should see what was ahead. I ceased dozing and looked. There was a herd of elk coming down off the bank onto the highway. We were just south of Humptulips a little way from the bridge. I thought it unusual that the herd was moving at night as they usually find a good place to bed down and stay there. There they were, though; something had started them moving. Later I had the chance to see what probably happened.

We watched the herd cross the road, a big herd led by a magnificent bull. Following him were cows and calves, and other bulls. Whether by accident or design, the last elk to cross was another big bull. The elk did not cross in single file, so we could not get an accurate count, but there were well over a hundred.

Under most circumstances elk are quite predictable. Game commissioners have told me that elk travel a circuit covering as large an area as 40 miles in diameter. Their arrival at a certain point in that circuit can be predicted within hours. They visit the same grazing and browsing areas each time around.

In one study of elk, the game commission demobilized two elk with darts, collaring each with a radio transmitter. One was released to continue with the herd. The other was "tightlined" down a hillside to a truck and hauled 30 miles away, where he was released. The two transmitters were monitored constantly. The elk released had immediately proceeded in the known orbit of the herd. The one released 30 miles from where he was captured did not return to that place. He headed for a point in the circuit where he would intercept the herd. In less than two days he had rejoined them.

The owner of one of the smaller logging companies said he knew I liked to hunt, and wondered if I would like to get an elk. I said I sure would. It was two days before the season was to open on elk. My friend said he thought he knew where they'd be the morning the season opened. He said they were due to come into his pasture along the river the next day, that they generally spent the night there and there was good feed so they wouldn't move too early in the morning. Something prevented my being there the morning the season opened. Several other men from Forks were at the spot at daylight that morning. They came into Forks later with three elk loaded on a pickup truck, with its springs mashed flat.

The largest remaining herds of *Wapiti* (Indian name for elk) are found in Yellowstone Park, on Sun River in Montana, and in the Olympic Mountains in Washington. Game commissions make a differentiation between so-called Rocky Mountain elk, and Roosevelt elk found along the Pacific Coast. Farther south, in Oregon and Northern California, the elk are much smaller than those in the Olympic Mountains. There the elk are as big as the Rocky Mountain breed; antlers which are shed to grow new ones each year can reach six feet across and have as many as twelve points on a mature bull.

Men working a logging show often saw a herd of elk passing along a hillside near where they were working. The noise of diesel engines running did not seem to frighten them, but they would keep their distance. Once up the Hoh River I happened to have a camera when a herd crossed the road. I stepped out to take a picture of them and a big bull, probably the dominant bull of the herd, squared away at me and lowered his head for a charge. I got the picture, but I also got back in the truck. I remembered the time one of my friends was charged by one bull while shooting at another. He got so excited he jammed his gun and took to a tree. The tree was small and the elk bashed it again and again trying to shake the man loose. He yelled for help, and it finally came. The elk was shot before it could dislodge the terrified hunter.

Hunting took us to some spectacular places. One long logging road climbed from the valley floor winding its way upward through cuts, around switchbacks, and across fills and bridges to the very top of the range of hills east of Forks. Loggers called this the Burma Road because of its resemblance to the original in Asia. Once on a cold clear day at the top we sat on stumps and looked a hundred miles out on the Pacific Ocean, over the tops of 30 miles of intervening hills. The only thing to mar the beauty of the magnificent view was the hundreds of square miles of logged off hills in every direction, except the southeast.

Deer on the Olympic Peninsula must have numbered in the millions. They were not easy to kill during the hunting season because there were hundreds of miles of heavy brush, and they would hide there within minutes after the first shots of the season were fired. Yet, because of their large numbers, most hunters got their deer before the season was over.

One morning, on the old Quillayute road, several of us were on the road above a logged-off flat bounded by timber on the other side. We all saw the buck about the same time. Guns started roaring. The buck continued, apparently untouched. Twice I thought I had the bounding deer dead in my sights. Just before escaping into the timber, he dropped. Everybody went down the hill and across the flat to where the buck had dropped in the rubble left from logging. We stood for a moment looking down at him. One of the men said he didn't know how we could decide whose buck it was, as we had all shot at him. I saw the glint of metal at the deer's shoulder. I reached

Blacktail deer of Olympic Peninsula. *Courtesy Washington Department of Game, Olympia, Washington.*

Olympic cougar. *Courtesy Washington Department of Game, Olympia, Washington.*

down and plucked the fragment loose. I asked if any of the others were using silver-tip ammunition. They all said no; it was my buck. At home, dressing him out, I found another slug. It too was a silver-tip. The fragment I had found sticking in the hide was part of one slug that had gone through the buck's heart breaking the skin on the other side.

Animals on the peninsula sometimes grow to unusual size. A predatory animal hunter came through Forks one day with a cougar wedged between the hood and fender of his old car. The cat was a female measuring 11 feet from tip to tip. He also had two live cubs. They were used later in a Disney film. Cougar were numerous there, because deer were plentiful, but they were not sighted as often as deer, elk and bear. They have an uncanny ability to stay out of sight, but we often found tracks only minutes old.

There was no bounty on bear, but the season was always open for them, except during the deer and elk seasons. Ranchers hated them because they kill calves. Others hated them because they kill elk calves. So, on weekends, when there was no work to do on our trucks we often hunted bear.

About dusk one evening, my son and I were near the bank of the Sol Duc River (original spelling). We had been finding tracks and signs all afternoon, now the tracks were so fresh that in the muddy places they were still filling with water, and on dry dirt the sides were still crumbling in. They were big tracks, and we were very close to him. We heard elk bugling ahead. We knew the big bear was as aware of the presence of the herd as we. We kept on toward the Sol Duc, hoping to see the bear before he killed an elk calf. We came to a clearing, and there was the elk herd. It was a large herd, ready to bed down for the night. It was getting dark rapidly, when the herd began to get very restless. Two or three bulls began to bugle. We were sure the bear had been detected by the elk. It was too late to shoot with accuracy, and we certainly didn't want a wounded bear of his apparent size on our hands. Then, to the right of the herd, we saw the bear. Indeed, we could hardly help seeing him. He stood on his hind legs for a moment, then dropped and rushed the elk. The elk began swarming toward us at a run. We jumped into the timber for safety, and the herd thundered by. They would find another spot

Olympic Elk. *Courtesy Washington Department of Game, Olympia, Washington.*

to bed down. We saw no more of the bear, and it was too dark to shoot, so we headed back to the car and went home.

Near that spot, a big clumsy boy had got a job with a crew of construction men. He was the water carrier. Going from the worksite down to the creek with his two buckets one day, the obese kid came around a stump and surprised a big black bear digging for roots and grubs. The lout was so shocked that he stumbled and fell, hard, right on top of the bear. Still spasmodically clutching his buckets, he ran to the creek as fast as he could make his fat legs go. Eventually he got himself in hand enough to fill his buckets. From far up the hill he began to hear faint cries of "Water Boy". He knew he had to return sometime, so he started back. Creeping up the trail with great caution, and overwhelming fear, he reached the place of the encounter. No bear. A little relieved, watching both sides furtively, he continued up the trail but far from over his fright. Then he saw the top of a black head and two eyes peering over a log. The half-smashed bear evidently didn't know what to make of an assailant who had the gall to pounce on him with all his weight, which was considerable, and the overweight water boy made it back to the job in an admirably short time. He never went there again. The boss had to recruit a new water boy.

The biggest black bear I ever saw we killed on the peninsula. Putting him on a scale at the local feed store, he hung off the scale on both sides. A substantial part of him was on the floor. The part that *was* on the scale weighed over 400 pounds. Two of us had had a gruelling time dragging the bear to the car. Every time we pulled on the rope we had around him, he stretched. When we stopped pulling, he slumped back half as far as we had pulled him. It took us an hour to get that monster the 100 yards from where we had dropped him to the car. Loading was even worse. We had to pass the rope around a trunk hinge to make any headway. Both of us would heave on him as hard as we could and one would catch the slack quickly before we lost it. We did get most of him in the trunk, and trussed the rest of him to the trunk hinges so he couldn't drag the ground and slide out.

The old bear was scarred from many a battle. His was a trophy hide. Word got around town and we soon had calls. One very rich woman wanted the hide to have made into a rug with the head on. She didn't bat an eye at the price, so we skinned the bear and delivered the hide to the tanners. The trunk hinges were never right after the abuse we had put them through that day.

I believe it could be established that in a frontier situation, where wildlife abounds, rivers run white and swift, mountains scrape the sky and timber stands tall, men respond to that environment by becoming a little tougher, a little more primitive and courageous. Part of the phenomenon may be that a country like that attracts that type of man. In any case, generally speaking, from Port Angeles to Aberdeen, men were mostly real men. A bit of the wildness of the animals seemed to be in them.

Willie had a touch of that wildness in him. Before he went into logging Willie had been a Golden Glove boxer. He had established quite a reputation, winning nearly every fight. He soon found himself with heavy responsibilities. His father had died, and there were brothers and sisters to be provided for. He turned to logging, where he could make a dependable living. Willie worked his way up rapidly. He soon established two facts: one, he was the best loader in the country; the other that it was a mistake to cross him.

By the time I came to Forks, Willie was running one of the hottest logging outfits on the northern part of the peninsula. He was known as a tough boss. Driving out the highway to his works, Willie would be calm and composed. As soon as he arrived at his logging show, even before he got out of his car, he started breathing fire. Things had better happen, happen right and happen fast when Willie was around, and the men all knew it.

Everybody gave Willie credit because, having taken charge of the company when it was having difficulties, he had put it in the black in short order. He was a good operator, and a shrewd timber buyer. When the price of cedar was high, Willie logged the patches that ran heavily to cedar. When cedar went down (it flucturated more than other woods), he logged the stands that ran mostly fir. Some spruce, cedar and hemlock were nearly always present in any of the stands, and of course they were all logged together, but Willie kept several areas of timber bought ahead so he could log where his company could profit most. He would get the local commercial pilot to fly him over areas that he believed to stand heavily in cedar. Flying low and using binoculars, he could make a fair estimate of how much cedar per acre he might expect. Then came the actual timber cruise, and finally the negotiations for purchase. Willie liked cedar for another reason when the price was right: almost all of a log, or a split log, or even a big chunk, was salable since it went mostly for shingles and shakes. It didn't take a very big slab of cedar to be usable for shingles. The haulers liked cedar, too. A big load was not excessively heavy, and the scale ran high.

Truckers hauling for the company that Willie ran were considered most fortunate. They seldom had to wait long for a load, and Willie taught his loaders to build a load almost as well as he could. When in the woods, Willie often took part in the loading himself. He was still the best loader in that country. But, like the country, Willie was wild. Any negotiations, whether with timber owners, or with his men, were carried on in as brief a dialogue as Willie could make it. He called a spade a spade, and took no guff; he would not tolerate subterfuge. He believed that when words would not suffice, fists were next in order. Willie had one slight handicap. He had an unusually high voice for a man; but so did Lincoln.

Large male Olympic black bear. *Courtesy Washington Department of Game, Olympia, Washington.*

When Willie wanted something, even if it was only a haircut, he wanted it at once. The barber in Forks was a rather strange guy. He had bought an old hearse for his own use. He had wild schemes for making money that he was eager to inflict on his victims while he had them in the chair. Willie didn't like the barber, but he knew the guy could cut hair, so he ignored his dislike to the extent that he settled for letting the man cut his hair, while at the same time he hung the nickname "squirrelly" on him. The barber didn't stick to his shop too religiously, and it was not uncommon to hear Willie out in the street demanding of the public "where's squirrelly?" in a shrill voice that could be heard above the sound of passing trucks.

Then one day when Willie was assisting with the loading, he got nastier and nastier with the puncher of the track loader. Willie accused him of clumsiness and incompetency, knowing the man was the best in the business. He cursed and stormed and in his expanding rage invited the man down from the loader so he could whip him. Paul, the loader puncher, accommodatingly dismounted, and the loading came to a halt while the issue was being settled. Willie told him to come and get it but Willie was the one who got it. They met in the mud by the log pile. With all the skill Willie possessed, he feinted, then aimed a blow at the side of Paul's head. The blow was blocked and Paul reciprocated with a solid blow to the body that knocked Willie back ten feet. Willie screamed for Paul to see if he could do that again. There was another quick exchange, and Willie was on his back. The next time he bored in hard. He put everything he had into the attack. Paul proceeded to put everything Willie had, including his face, in the mud. I never saw a man harder hit. He had turned a complete flip in the air, and came down on his face and chest. Willie didn't get up right away. Paul climbed upon the loader and prepared to resume work, if Willie got himself out of the way. Willie stirred, sat up, got to his feet, and gazed around. Something was wrong here. He, Willie, was supposed to be the expert fighter. No, something had definitely gone wrong. it had been some kind of fluke. Willie turned toward the loader and stared. He said he guessed maybe they better try that again. Paul said it was OK if Willie was sure that was what he wanted. He spoke in a most gentle kind of way. There was an almost exact repetition of the first engagement, only this time Willie went down three times before a final blow turned him over in the air and he ricocheted off a log, and lay still.

The victor once again mounted the loader and got ready to start the loading again. Willie roused, picked himself up out of the mud, and grabbed a pair of tongs. The interlude had come to an end. The incident was closed. Just as George Raymond had done, Willie overlooked the whole affair. He would never have considered firing a good loader puncher just because he had worked him over.

I noticed one great difference between the men of that time and those of a few decades earlier. None of the men around Forks would ever have used the "corks" on a man once they had him down. Those many years before, it would have been standard procedure, but not by all men; some men had always had too much real manhood to take advantage of a man who was down to scar him for life.

CHAPTER 19
RIGHT SIDE UP

One day, while waiting for a load, south of Forks, I heard a comment that chilled my blood. The story really begins when I had left the body shop to go back into trucking. My son, Dan, soon had the offer of a different job, too. He began driving truck for one of the major oil distributors in the valley. A major part of his job was to deliver diesel oil to the various loggers in the surrounding country for use in their yarders, cats, and loaders. There was still that one outfit using an old steam donkey. For that Dan delivered fuel oil. In the course of his deliveries, Dan had to pull a fully loaded tank truck *up* a number of hills that loaded log trucks came *down.* The roads were not really designed for loads to be taken uphill. One of the roads was the long hill up to the L.O.G. Company's works. Another was the climbing, twisting Burma Road mentioned earlier. Loaded, Dan did not dare to stop on the hill. It was all that Mack could do to *keep* the load moving. It would never have started again if stopped on some of those hills. Trucks coming down sometimes couldn't stop immediately. There were some hairy experiences. Dan tried to take his loads to the worst places earlier than the first load would be coming out, but it wasn't always possible. In any case, even if not on one of the hills, there were always times when he would meet a load of logs unexpectedly. An empty log truck returning could stop quickly and back up to accommodate the loaded one coming out, but the loaded fuel truck did not stop so readily, and was harder to maneuver back off the road.

Another stretch that worried me so I had asked Dan to be especially careful on it was the Clallam Bay road. The road was narrow and crooked. It wound its way from the inland valley through mountains where all the turns were "blind", to the shores of the Strait of Juan DeFuca. While I was waiting at the landing for a load, one of the incoming drivers asked another, as he crawled out of his rig, if he had heard about the oil truck that went over the bank that morning. The other driver answered no, and wanted to know what happened. The first man said that was about all he knew, really, that he had stopped in town and somebody was talking about it. I stood still while the others conversed, wondering if it could be Dan, wondering if he was hurt, or even dead. I got hold of myself and asked the driver if he had heard what had happened to the driver, and he had not.

I loaded and headed for town, and drove around to the hospital. I had to find out, and I did. It was Dan's truck all right. I learned that the truck was said to be a total loss. Dan had been brought in to the hospital where he had been checked over and released. I drove to where Dan lived, just two houses away from mine. He filled me in on the story.

Loaded for deliveries at Clallam Bay, he left the valley and went into the hills where the road was narrow and winding. At one point the logged-off land went up steeply to the left and down sharply to the right to a marsh 300 feet below. Dan had pulled into a turn to the left that was blind, and around the turn from the other direction came a car pulling a trailer house. It was in the middle of the road. Dan knew if he hit the car, or the trailer, with the corner of his tank truck, somebody would be hurt or killed. Neither of the drivers could possibly stop in the brief moment of time before impact would occur. Dan had to make a split second decision. There wasn't room to pass the car and trailer without going to the edge of the shoulder. If the shoulder was solid, and the sidewise surge of the tanker didn't throw him over the bank while swerving around the oncoming vehicle, he might get away with it. He cranked the steering wheel hard right to get over, then hard left to recover. The shoulder caved just as he was clearing the trailer. The outward surge of the heavily loaded tanker, as Dan wheeled it to the left, trying desperately to stay on the road, gave impetus to the roll as the truck went over. When he saw that he had lost it, Dan had one fleeting impulse

Approaching the Elwha Bridge from the west on the way to Port Angeles. Truck on bridge running west empty. *Courtesy Paul Deines, Port Angeles, Washington.*

to bail out, but the tanker load of fuel revolved so rapidly over and over, that there wasn't a chance. The left door came open, and Dan didn't know whether he instinctively grabbed the handle or not. He didn't know anything, except that he was taking a terrific beating in the tumbling, pounding truck that was likely to catch fire and burn. The time the truck spent in its wild roll couldn't have actually been long, but to him it seemed like the thing would never stop. The door had been torn away as soon as it flew open but for some reason Dan wasn't thrown from the truck.

The truck came to rest with the open hole, where there had been a door, toward the sky. Dan scrambled out, gasoline was pouring from the tanks, and the filler caps had been knocked off as the truck rolled over stumps. Certain that the truck was going to burn, he scrambled to safety and made his painful way up the long, steep, stump-cluttered hill to the road above. He hurt all over. A car came along and the driver saw the situation. He gave Dan a lift to the Forks Hospital. That was about it. The old Mack was all through hauling fuel to the loggers. Fully insured, there was no financial loss to the distributor. Dan continued his work in a day or two, still sore all over, but he did it with a new truck. There was one other thing, the driver of the car and trailer that crowded Dan off the road didn't even stop.

About that time I was having troubles of my own. I had bought a Kenworth, intending to transfer the White to a shingle haul as soon as the deal was approved. In the meantime, I hired a driver to haul logs with the White. Within a week, I had lost a differential in each of the trucks. Each rear-end meant a $450.00 replacement plus two days of lost hauling, another $300.00. The two incidents therefore cost me $1500.00. The White needed other repairs, too, nothing major, but the driveline was wobbling and vibrating and we had tried several times to correct it without success. The truck needed a new shaft. The haul to Port Angeles was a hard one at that time, because at least five miles of the road was under construction. The road was composed of a heavy,

Load of logs on Elwha Bridge. *Courtesy Paul Deines, Port Angeles, Washington.*

gummy combination of mud and gravel. Cats were kept available on that stretch to pull trucks that were either stuck or broken down. Many "bogies" (rear-ends) went out during those months. The road just pulled the guts out of trucks.

While I was waiting to recover from the loss of the two "bogies" I had put off installing the driveshaft in the White, but I waited one day too long. Coming down into the Elwha Valley from the west, I was surprised to see my White midway on the hill, facing me with its trailer cocked sidewise on top. The driver was standing by the truck. He looked like he was in shock. Pulling the Kenworth over to the left on the shoulder, facing the White, I got out to see what was wrong. I saw plenty as soon as I looked at the truck. I asked Ron where the driveline was and he said he had lost it at the top of the hill. Puzzled, I wanted to know if it was that same hill we were on. He said no that it was on the hill on the other side. Completely baffled, I didn't see how he got as far up the west hill as where we were, he stated he had "coasted". I couldn't help saying that was impossible; I thought he had lost his mind. The top of the other hill was two-and-a-half miles away, and even if he had coasted down that, how in the world did he get a quarter of the way up the hill where he was now? He said all he knew was that just after he came over the top of the other hill he heard a bang and the truck sort of jumped, and the vibration stopped. He looked in the mirror and saw the drive shaft and some other pieces bouncing down the road. He tried to put on the brakes and the pedal just hit the floor. So he didn't have any brakes or gears. The truck just kept picking up speed all the way down. I looked at the truck again, and saw that the drive shaft had hit the brake line when it carried away, letting all the fluid out. No wonder Ron had no brakes. I asked why he didn't bail out at the top and let the truck go over the bank. He didn't know, but he just couldn't bring himself to do that. I said he *must* have known there wasn't a chance in a *million* he could ride it out in one piece. But he said anyhow he didn't jump and there she was.

Ron told me that as the truck gained speed down the long hill, he regretted that he hadn't jumped. He said that before he reached the last turn, just before entering the Elwha River bridge, he felt that he was almost flying, moving faster than he had ever done in his life. He said when he got to the turn at the bottom, where the State had graded it out extremely wide, he had used it all. He had laid the truck over to the left, almost against the bank, and then wheeled it to the right as hard as he could, trying to make the bridge. He said it was strange that it was the first time since he had been on that road that there wasn't another truck, either on the turn or on the bridge, that the whole truck had broadsided into the bridge and slapped the left railing *hard.* Afterward he was surprised to find the trailer still on the truck. He thought he had dumped it into the Elwha.

Then we heard a viewer's version of the incident at the store across the bridge. The manager of the store exclaimed that indeed he had seen it, that the truck had to be going a hundred miles an hour when it slid around that corner. He thought the whole thing was going right into the Elwha, he never thought it could make the bridge. I asked if he thought the driver had gone crazy, and he said no, he knew very well nobody would come down that hill like that on purpose, that it had to be a runaway.

I, like Ron, had been hauling over that road for many months, along with several hundred other trucks and thousands of assorted cars, buses, and other vehicles. The road was always heavily travelled. Never once in all that time had I seen both the bridge and the curve without traffic at the same time. Sometimes it is hard to believe that there are not guardian angels taking an occasional hand in the affairs of men.

That night we installed a new driveshaft and two universals, and rebuilt the brake-line system. The White went down the road on to Forks without a bit of vibration. It operated better than it did the day I had bought it.

Truck loggers have an old expression they often say to each other just before a trip: "Keep 'er right side up". Ron had managed to do that, just barely, but Dan had not.

Looking west toward the east entrance to the bridge over the Elwha River and the turn around which the White ran at an estimated 100 miles an hour. It was from the building on right that the speeding truck was viewed. An empty truck with trailer can be seen going west on bridge. *Courtsey Paul Deines, Port Angeles, Washington.*

CHAPTER 20
SITKUM OPOOTS

After a couple of years there came a lull in logging. Stumpage that ran 20% or 25% peeler (large smooth logs that could be cut in thin layers for plywood) was selling for $110 and $120 per thousand board feet. Competition for stands of timber that could be logged profitably was running very high. Whenever a sizeable stand of timber was advertized, all the big companies bid on it, and a lot of the smaller ones did too. Companies that had been counting on cedar for their best profits found the market for that wood down and some of them ceased operations temporarily because the stands they were logging ran heavily to cedar.

Suddenly there were a great many more trucks needing jobs than there were jobs needing trucks. By that time I had gone into partnership with Lenny Nelson. We both found outselves temporarily out of work because the company we hauled for had shut down "until further notice".

We would have headed south in search of hauling, except that truckers from Oregon had already arrived on the peninsula looking for work. We were not too worried, though; we knew things would pick up again.

In conference with our drivers, we considered the idea of finding small, isolated patches of timber that the big outfits couldn't afford to move into, and logging them ourselves. Then, we reasoned, we wouldn't have all our eggs in one basket. If we didn't make much hauling the stuff, we might show enough profit in the logging to come out all right anyway. We all wanted to try it.

As our first move, we went to Mount Vernon, Washington, and bought an old Cletrac, a few blocks, some tongs and enough line and other equipment to get by with for a short time. Later we would get a larger cat, a logging arch and a loading donkey.

As soon as we were ready to operate, one of the Forks loggers decided we needed a catchy name for our outfit. He declared it should be called the Sitkum Opoots Logging Company. In the Chinook Indian language, sitkum means half, and opoots means posterior. We didn't call our outfit that, but our friends did, so we finally gave up and identified with the rather amusing pseudonym.

As our first project we moved to Nisqually, where I had some acreage standing thick with small timber. It was too small to pay to cut for mill logs; however, there was a good market for piling, and for the really small trees there was a market in Japan. They use smaller piling in Japan than we do here, so long as it is long and straight, and that stuff surely was.

We camped in the timber and made our own breakfasts and lunches, but went to a restaurant for a decent meal in the evenings. Falling, trimming and topping that timber was just as tiring as the work in big timber, but the number of "sticks" we could produce in a day was quite different. We went through that stuff like a bull through a corn patch. Because the stand of timber was so thick, we had so many on the ground, trimmed and cut to length that we could hardly log them out. We had to get a long load permit and the required insurance, of course. We were not only paid by the linear foot, according to size, but the longer the piling was, the more we were paid per linear foot, so we tried to get the maximum length out of each tree.

The little Cletrac was equipped with a blade so the first thing we did was cut a strip the full length of the property which we bladed out for our road. We erected a two-guy gin pole at each landing as we came to it. Opposite the gin pole we always left two tall stumps to serve as a brow log. The tongs had to raise each pile higher than the truck before it would come over the truck. When there was enough piling on the landing to make a load, the Cletrac did duty as a loader. When there was not, the Cletrac was our yarder. When it threw a track, which it frequently did, it became a pain in the neck. My partner and I rarely agreed on anything, but our quarrels never were very belligerent. We were contented to settle for insults in place of blows. One of us would ask the other where he learned to log, pointing out that he was a knothead. We spent every day working, sweating, breathing hard and insulting one another's intelligence quotient. Yet, when it finally came time to go to the restaurant for our evening meal, likely nine o'clock, we laughed and joked all through the meal. Once in awhile one of our friends from Forks would drop by on the way to Seattle. If he wanted to spend the night, he was welcome. We had plenty of room on the ground where we slept.

Because of the long reaches we had to have, our trucks were overlength, even when empty. We couldn't even load the trailers for the trip back to the woods because of the length of the reaches. We graded the entrance of our road fifty feet wide so we could get out onto the road without the trailer wheels going into the ditch. Even then, the truck had to cross the road completely, turning only at the very edge, and blocking the road before turning.

The piling went into Tacoma for shipment to Japan. It was only a 30-mile haul. I could remember hauling where a 30-mile haul would have been considered a long haul indeed. But the distances we had hauled into Port Angeles and Aberdeen made this seem a short haul. The trucks had no trouble, and no tickets. That was a good thing, because we had plenty of trouble in the woods. The chain saw would have to be sharpened. One of us would have to go to Olympia for a piece of equipment. The Cletrac would throw a track. The fire warden would warn us about the humidity. Then somebody would come to see if we were cleaning up our brush, and the Cletrac would throw another track. *That* we could depend upon. Everytime we got that crawling thing in a bind while yarding a log out of an awkward spot, or got it on a sidehill, or turned it too sharp, the thing would throw a track.

Everybody was glad to finish the piling job and go back to hauling for other companies. We began pounding the highways again between Forks and Port Angeles, or between the Hoh River and Aberdeen. We hauled steadily for months, and got in pretty good financial shape. Yet, we had had a taste of operating on our own, and we constantly looked for timber on days off and evenings. We found several patches but only three that we considered worth logging. Two of them were in the Grays Harbor country and one was near Kapowsin. That one was a respectable stand of old growth Douglas fir. It had been left by some company in the early days because it was high on a bluff, too far from their railroad and too hard to get out. Later roads had been built, and one abandoned. From the old abandoned county road we could get the timber off the bluff and load it at the old road. A cat would log to the bluff and a donkey would both yard and load.

When we got ready to start we had an RD-9 cat, a track arch, plenty of rigging, and the promise of a donkey. We couldn't get the donkey until its owner had finished about 60 days work with it. While we waited for the donkey, we decided to go ahead on the other two patches, so we moved south of Grays Harbor.

While some falling and bucking was going on, I bladed out our road and even took one or two small grading jobs with the cat. Then before we rigged our tree there, we had a "cold deck" (logs cut and stacked) to load out, five miles south. It was going to be a problem to find a good place to load our cat. I asked Len why load it and he pointed out that it was against the law to "walk" it on the highway. I told him I'd walk it down the beach. He said it was private land between there and the beach; anyway he didn't know if I could get across the drift logs. I told him to get the rigging down there and I'd be there by the time they needed me.

I knew nobody was going to object to my walking the cat from the road to the beach. How much can you hurt sand dunes? We planked the cat across the highway and I headed across the dunes. When I got to the drift logs, I shoved some out of the way, I broke some in two, and I climbed over the rest, standing the old RD-9 on end a few times. Then I clattered on down the beach for about an hour, and went through the procedure again, getting up to the cold deck.

The boys had just unloaded the rigging. Belted and spurred, I went up the tree, still standing, that had been used to stack the logs. I pulled a small line up after me, then a light block which I hung so we could pull the heavy gear up and rig the tree for loading. We rigged a crotchline outfit and I came down from the tree. We were ready to load. We dragged one of the bigger logs parallel to the truck road for a brow log and our first load was soon on its way to town. We loaded out and hauled that cold deck in about a week. The logs had been stacked for a couple of years, so there was some loss from sap rot and worms, but it wasn't too bad.

We felt good about the operation because we had proved to ourselves that we could rig for an operation quickly enough to be able to go after the small stands that most of the logging companies would leave because of the cost of getting the timber out.

We went to our regular show. Looking the situation over we found that there was only one place ideal for the first loading tree, as the timber was in swampy country, but there was no standing tree there suitable for a spar tree. We didn't consider that to be much of a problem. Finding a suitable tree, we cut and trimmed it, and dragged it to where we wanted it. It was placed with the butt against a stump, so when we pulled on it from the fairlead of the arch, which was eight feet high, it would upend.

Four hours were spent in measuring, cutting and splicing guy lines. Another half-day was spent securing the lines to the tree on one end, and stumps on the other. We wanted our tree to lean out over the truck road as a big gin pole, hanging into the two back-guys. But there was a problem. The tree had to be prevented from swinging sidewise as it was being erected. It was concerning this matter that my partner and I had one of our more serious differences. Lenny didn't want to "waste the time" putting control guys on the tree, secured to stumps that were at opposing right angles to the tree's butt. He told me to just get on the cat and pull it up. He thought he could hold it with a rope. I said he couldn't begin to hold that tree with that rope, and anyway what if it should swing toward him, what would he do then, *push* on the rope? Lenny was in a hurry to get to logging and he wouldn't listen. After ten minutes of verbal futility, I offered a compromise. I would try it on one condition, if anything went wrong and in case he was still alive we'd do it the right way next time. Lenny agreed. I jockeyed the cat and arch into dead center position and lashed the cat to a stump. We hooked onto the spar tree two-thirds of the way up and I started to drum in the line. The tree snugged up tightly against the stump as the power was applied. The line got so tight it was spraying fine mist, as it had been wet. Nothing happened at first, but out of the corner of my eye I could see that Lenny was pulling hard on the rope. To my surprise, the top of the tree came 25 feet off the ground and hadn't swung. Then slowly like a ship beginning to drift away from a pier, the tree started to swing toward Lenny. There wasn't a thing he could do but drop the rope and run for his life. The trouble was as the tree pivoted and came crashing to the ground, smashing into other standing trees too small to stand the weight and momentum of the big tree, the little ones broke and fell. It was one of the little trees that came near to getting Lenny. In his frightened haste he tripped over a small log and fell flat. He was still flat when one of the little trees fell on top of him. The log that had caused him to fall had saved his life when the falling tree struck it just above Lenny's outstretched back.

I was satisfied Lenny had not been hurt, all we were out was some time. Taking the anchor line from the cat I shoved the tree back to the exact position it was in before it swung and nearly killed Lenny. We went about the business of putting on control guys. As I drummed the line in next time, the tree came off the ground, as before, but the guys kept it from swinging either way. Up and up it came until it seemed to poke a hole in the sky. As it reached an upright position, and stood tottering, I wondered if all four ends of those back guys were really secure, and if the lines would stay fast on their stumps or flip off when the strain came. The tree stood balanced, then began falling towards me. Its fall accelerated, the back guys snapped tight, the tree gave a couple of bounces and settled securely in the cradle of the two guys. Since we had completely rigged the tree before we raised it, we were ready to log.

Just as the tree swayed and dropped into its guys, old Fritz came wandering in. The timber belonged to him. In fact, several parcels of land around there belonged to Fritz, some of it timber land. Other loggers had refused to buy timber from Fritz because he wanted more for the stumpage than the coastal timber was worth, and they had heard stories that tended to discourage faith in his honesty.

We had figured out how to work with Fritz. He got the going rate for net scale at the mill. He got his percentage directly from the mill in South Aberdeen. If he was unhappy about his pay, it was between him and the mill. Whenever we did any grading or landfilling for him, it was for a stipulated price instead of by the hour, payable at the completion of the job. If he came around the logging works too often with his ideas and suggestions, we ran him off. The timber at that location was not too profitable. One, it tapered too much which is typical of much of the timber too close to the ocean. Two, it stood in a swamp where fallers and buckers often worked knee-deep in water, and setting chokers often involved dragging mainline out into the swamp to reach the logs because the cat had to stay on solid ground. Three, because of the nature of the timber, a mixture of spruce and hemlock, we could not get enough scale on a truck to make a good payload. But it was something we could do to make a living while we waited for our donkey. Then we could move into the stand of old growth near Kapowsin. Our prospects looked good, and we felt we had reason for optimism.

Then something happened that was disastrous, as far as I was concerned. I was taken to the hospital with a back injury. After ten days in the hospital I was released to go home with orders to stay off my feet entirely. I was also told that resumption of logging, driving cat or driving truck was out of the question, When weeks had gone by and I still couldn't walk without help, I decided to sell out. I sold my half-interest to a mutual friend of Len and myself. We had a dinner party at my house that evening, and I said goodby with some regrets to the Sitkum Opoots Logging Company (registered officially as *Nelson and Pierre*) forever. I doubted that I would ever log again.

My doubts and those of the doctors about the future condition of my back were unfounded. Doctors can be mistaken, particularly about prognoses. I learned that it is about as likely for a logger to turn permanently from logging as it is for an alcoholic to give up drink, a race horse to give up running, or a compulsive gambler to give up betting.

Three-horse team and load of shingles. *Sketch by Joseph H. Pierre, Jr., Monterey, Calif.*

CHAPTER 21
SAWMILLS

Before further discussion of the logging industry, something should be said about the plants that converted the raw log to a great variety of useful building products. After all, it has been many decades since houses, barns, and other buildings have been constructed of logs. The pioneers did it because of the absence of mills in those early days. Log houses had their good points, whoever heard of a log house blowing away in a wind storm? But it has always been more efficient, more practical and conservative of timber to build houses of lumber. Therefore, as fast as mills could be constructed in the Northwest a ready market was waiting for their products.

In the early years in the Northwest, mill machinery had to be shipped from the East either by train or by ship. The machinery for Pope and Talbot's first mill at Port Gamble came around the Horn in the schooner, *Foster,* which was "loaded to the decks to the water with engine, boilers, mill machinery, merchandise for the store and general supplies", (to quote Newell and Williamson in *Pacific Lumber Ships*).

Many of the early mills were unpretentious affairs powered by either an "undershot" or "overshot" water wheel, depending upon the "head of water" available. Some of these, called *sash mills,* used an up-and-down motion saw with a blade ten to twelve inches wide and ten feet long. A passage from W.E. Lamm's book, *Lumbering in Klamath,* reads "another old timer of Bonanza, when questioned about the Gordon Mill remarked, 'Yes, I knew the Gordon Mill. It was one of those up-and-down mills, up today and down tomorrow. Grandpap used to start the saw in the log, then go away sometimes to catch a fish, and then after awhile he would go back to see what effect the saw had upon the log!'"

Even mills using steam power and circular saws rarely cut more than 1500 to 2000 board feet per day, but there were enough of those little mills scattered around by 1881 that local needs were met. Mills that cut enough lumber to ship, such as the Henry Yesler mill in Seattle and the Pope and Talbot mill at Port Gamble, employed several men. Later both these mills and other high-producing mills, went to the use of band saws. To gain a substantial production of lumber an edger was needed. To further increase production, resaws were installed. The resaw reduced the work of the headrig. Where gang saws were employed, the "headrig" (primary saw) had only to square the log and cut it into cants of the desired thickness for the gang saws to reduce it to the lumber dimensions most needed.

Eventually planing mills made their appearance. These mills could either put smooth sides and edges on the lumber or, fitted with proper heads, produce tongue-and-groove flooring, and ceiling even decorated with a bead. Dry kilns and dry sheds were

Algoma Lumber Company in 1914, located in Klamath Lake which served as an ideal millpond. *Courtesy Bob Lewis, Klamath Falls, Oregon.*

International truck unloading at the water, from whence the logs would be towed to the mill. *Courtesy Bob Lewis, Klamath Falls, Oregon.*

When the body of water, on which the mill was located, was very large, as is Klamath Lake, logs often travelled the last ten or fifteen miles towed in a raft to the mill destination. *Courtesy Bob Lewis, Klamath Falls, Oregon.*

Lamm Lumber Company mill in 1914. Here too Klamath Lake serves as the millpond. *Courtesy Bob Lewis, Klamath Falls, Oregon.*

added and lumber was not planed until it had been dried by one of these means. Then came furniture factories, sash-and-door factories,box factories, and cabinet factories.

By the end of the first quarter of the 20th century there were huge mills running 24 hours a day, and cutting more than 1,000,000 feet of lumber each day. Some far exceeded even that. The Longbell mill at Longview, Washington, produced more than 1,000,000 board feet each shift, or more than 3,000,000 feet each 24 hours. Mills such as this were actually multiple mills, having several headrigs instead of one, and enough gangs, planers and other backup machinery to handle the huge output.

Lumber was handled a board at a time only on the green chains, and dry chains where grading had to be done. Otherwise it was moved by chains, live rollers, and lumber carriers. There were two prominent manufacturers of lumber carriers, Ross and Gerlinger. Both makes were similiar in that they could straddle large loads stacked on carrier blocks, pick them up and transport them anywhere in the confines of the general plant. Every mill, even a small one, had to

A log that filled the mill. Nine feet 2 inches. It was a big project for a small, old-time mill to reduce a log like this to lumber. *Courtesy of Lila Boge, Tillamook, Oregon.*

Tall lumber stacks in the storage yard adjacent to a mill. This lumber will go to various parts of the country by rail, some to other parts of the world by ship. *Courtesy Bob Lewis, Klamath Falls, Oregon.*

Early log flume was flooded to float logs to mill. *Courtesy University of Washington Library, Kinsey Collection.*

have its millwright, (a master mechanic who could keep all the machinery in repair and adjustment). In the big mills a crew of millwrights were needed.

Let's follow a log from a mill pond, where it has been dumped from a log train or truck, or towed in a raft by a tugboat. A skilled pond man, jumping from log to log with his pike pole, works the log to the end of the chute which describes a gentle arc from the deck of the mill down into the pond. As the end of the log is eased into the chute, a bull chain with its huge dogs catches it and the log begins its placid journey up to the deck of the mill. As the log is propelled to the deck the bull chain disappears through the floor, constantly travelling over two huge sprockets, one at the top and one at the bottom, to bring up log after log except when it is stopped to allow time to get logs on the deck out of the way.

When the carriage, driven by cables on drums, is finished with the log it has been carrying, the next log is forcibly thrown onto the carriage by rams called niggers that come up out of the floor by steam pressure. The log is dogged to the carriage by a man called a "ratchet setter" who rides the carriage back and forth. The log is set out on the carriage to the proper position to be slabbed. On a large log, the slab itself will contain considerable lumber which will be salvaged by the edger and the resaw. The mills we are speaking of have huge band saws that travel around upper and lower pulleys, ripping through the length of a log in seconds. At the end of the cut, the carriage travels quickly back and the sawyer gives a hand signal to the ratchet setter indicating what thickness he wants: four, six, eight, ten or twelve inches. Those big mills had "shotgun" carriages with big steam rams at both ends of the run, which the carriage strikes, activating each ram so it stops the carriage and starts it in the opposite direction abruptly. A ratchet setter can only ride such a carriage for a limited time because of the shock at each end of the carriage run. (Modern mills have electronic controls so the sawyer can maneuver the log for cant thickness, eliminating the need for a ratchet setter.)

The cant goes from the carriage to the edger, where the edges are squared, then on to another set of moving chains to the gang saws where it is no longer a huge slab of timber as much as one foot thick, but is reduced to 20 to 40 1 × 12 boards, or whatever the mill storage had need of. After air drying for the necessary length of time, most of the lumber will go through the planing mill, perhaps even having been resawn into narrower widths.

When the time comes for shipment, the lumber is loaded on a railroad car, or truck to go possibly 2500 miles where it will again be put in dry sheds at a retail yard for sale and delivery to some builder.

Other methods, some unique, have been used. Some logs travelled to the mill by flume covering several miles in the process. Other mills cut rough lumber and cants at a rough mill far in the hills and the rough lumber travelled by flume to a mill located on the railroad far down the mountain. One such flume was used by the Bridal Veil Lumber Company up the Columbia. Another, still operating at this date, is the Broughton Lumber Company, south of White Salmon, also on the Columbia River. This company has two mills; one nine miles up in the mountains where rough lumber and cants are cut. This rough material travels the length of the nine-mile flume in 55 minutes, at times floating slowly, and sometimes in the steeper sections going exceedingly fast. The Broughton mill at the bottom is equipped with dry sheds, gang saws, a planer mill and everything else needed to make any kind of lumber the market might require. The lower mill is located on the railroad.

Many of the mills on the large rivers or bays or on the Sound, have trucks dump into a booming ground from which rafts are towed to the boomed off area used as the mill pond. Also, more and more, dry land storage of logs is used, logs being moved to the mill from there as needed.

There are so many products derived from the log that it almost boggles the mind: plyboard, particle board, chemicals, cork, mulch, some kinds of cloth, shoe polish, floor polishes, carbon paper, cosmetics, bulletin boards, paneling, gaskets, glue extender,

Broughton Lumber Company mill from the air. Columbia River at lower left. *Courtesy Broughton Lumber Company, west of White Salmon, Washington.*

Flume at Broughton's upper mill, lumber moves placidly at first.

Wigwam burner used to burn waste at Broughton's upper mill.

Broughton's upper mill far up the mountain from the Columbia River.

Moving placidly again, approaching the lower mill, nine miles from where the lumber started.

Flume at Broughton's lower mill, racing lumber on steep portion of flume.

Longbell Mill, Longview, Washington. *Courtesy International Paper Company, Longview, Washington.*

plywood adhesives, plastic filler, a phenol substitute, and the miracle drug DMSO (Dimethol Sulfoxide). Slabs and bark that used to go into the burner are now processed to make useful products, many not mentioned here. As time goes on, less and less of the log is wasted. Already nearly all of it is used for something.

In the early days, mill towns grew up spontaneously, and haphazardly around the mill. Later, some of the towns were planned at the time of the mill's construction. Such was the case with Longview, Washington. The city was planned completely before the huge mill, at the mouth of the Cowlitz River, was constructed. Around the big plant thousands of feet square, on two sides, the city has broad streets, adequate utilities and might be called a model city. The Longbell mill later was bought by International Paper Company.

Skidroad with horses pulling sled load of shingle bolts. Mason County. *Courtesy Custom Photo Company, Littlerock, Washington.*

One type of mill that has been around almost as long as any is the shingle mill. Shingles have always been in demand for roofs, and sometimes for siding because of better weatherproofing and additional insulation. Until early in the 20th century the shingles were always made of cedar.

Logging for shingle mills was generally done differently than for sawmills that produced lumber and timbers. For many decades, when a cedar tree was cut down, instead of leaving the woods in the form of a log, it was cut in short lengths called *shingle bolts.* These went from the woods to the mill on large sleds, over skidroads. Oxen were commonly used to pull the sleds at first, later horses were used, though again there was no point in time when this transition took place. Both were used at the same time by different operators and were sometimes even used together in the same team.

There were no giant shingle mills. The bigger ones might have had as many as a dozen shingle saws, mechanically designed to cut one shingle tapered one way and the next tapered the opposite way.

My relatives, the Brown and Sturdevant Company, produced shingle bolts for several of these mills, in Thurston County, Washington. Clark and Peterson operated a shingle mill in Snohomish County. Chris Peterson supervised all the woods work, including the falling, bucking and transporting over the long skidroad by horses. That old Dane, new to this country though he was, knew every tree that grew in the Northwest by sight. More canny, shrewd and conservative than some members of the firm. Chris accumulated considerable money from his part in the operation but that profitable venture came to an end quite abruptly one morning. The dry kiln burst into flames; cedar burns fast and hot. The entire mill was consumed within a matter of minutes. There was no insurance and some of the participants in the venture were left virtually destitute. Chris, on the other hand, was not without resources. Using the money he had saved he purchased a farm in Oregon and became a successful grower of filberts. He planted the trees himself and did general farming until the large orchard reached maturity and yielded thousands of pounds of nuts each year. He died independent at age 93.

Shingle sawing was a special skill requiring a man who had learned, often the hard way, to move fast. Most shingle sawyers ended up with less than a full complement of fingers. The shingle weaver worked below the sawyer where his shingles came down to

Six-horse team taking load of shingle bolts to mill.

him through a chute. He too had to be skilled and fast. As the name "weaver" implies, he weaved the shingles into the familiar bundles, butts out, securing them with an upper and lower cedar strip nailed together through a narrow metal band.

Perhaps shingle mills in general have had fewer innovative changes than lumber mills, small, medium, and large sized shingle mills dotted the Northwest before the turn of the century. They still do. The market for their products is still strong. Composition shingles and mill-cut shakes have been available and widely used for decades. But the old-fashioned cedar shingles, available in standard and long lengths, are widely popular, relatively inexpensive and durable. They will probably be around for a long time.

The Coats Mill in Tillamook, Oregon. *Courtesy John Coats, Portland, Oregon.*

Tethered balloon, used to lift logs high above the ground, was pulled out to the cut timber and back to the landing by lines from a diesel yarder at the landing. *Courtesy Bohemia Corporation, Eugene, Oregon.*

CHAPTER 22
LOGS THAT FLY

Many were the changes in logging methods between the time we logged the Tillamook and Astoria areas, and the time in the fifties when logging was going strong on the Olympic Peninsula. But there were more spectacular changes to come.

In the fifties, fallers and buckers were already using power saws. Most of these were powered by gasoline engines, but some were electric. The electric generator, driven by a gasoline engine, was mounted either on wheels or on a sled so it could be pulled to each new location by a cat. These electric saws required hundreds of feet of electric cable so a large number of trees could be reached before the power unit had to be moved. The gasoline-powered saws were often two-man saws where big timber was being cut. There were two reasons for this: one, it would be extremely difficult for one man to effectively handle a saw that long and heavy; two, it was wise to have a man on each side of the tree being felled so when the tree was ready to drop, exactly the proper amount of wood would remain holding on each side, and the tree would fall where it was intended to go.

Some of the "cats" used in that period were as big as D-9s and TD-24s and the logging arches were bigger and tougher than those used in earlier "cat logging shows." The donkeys would range from 200 to 450 horsepower, depending on whether the donkey was used for loading, yarding, skyline or slackline work. Much of the loading was done with a specially engineered track loader.

In case a spar tree was used, it was a standing tree that was limbed, topped and rigged, just like in the old days. If a growing tree did not happen to be where one was needed, one would be cut, trimmed, and dragged to that spot and raised.

Loading systems still included the simple crotchline, the crotchline with a spreader bar, the hayrack boom using two pairs of tongs, and the heel boom with a single pair. Hard hats were rarely seen in the woods except in a few of the larger operations where the safety engineers had ordered their use. Fatalities were reduced radically from what they had been in the highball days early in the century, but we *got logs*. Every man from the timber cruiser to the pond man produced a day's work, using the best equipment available to us at that time. Trucks were mostly GMCs, Kenworths, Internationals, and Macks. No one then visualized the changes that would occur in the next couple of decades.

A visit to a logging operation today will reveal some dramatic changes, and some not so dramatic. The use of balloons and helicopters for yarding is becoming more and more common.

The use of the helicopter for yarding can be simply explained. The helicopter hovers by the logs already fitted with chokers while they are being coupled up (unless grapples are used), then maneuvers to the landing where the logs are to be loaded, hovers again while the chokers are uncoupled and return chokers are secured, then flies the chokers back to the woods where the routine begins all over again. Information from the University of Washington and other reliable sources indicates that, because of the high cost of helicopters, and the inherent danger of men spending every working day of the month, rain, blow or shine, hovering near the ground, half the time heavily loaded, helicopters are not likely to come into general use very soon. The forestry services use them on a limited basis, and some large logging companies are beginning to experiment with them.

Balloon logging is coming into more common use. It is much less dangerous, much less expensive, and more practical. But from the rigger's standpoint it is more complicated. It is impossible for the balloon to provide more than one thing, lift. All maneuvers, all elevation must be impelled and controlled through lines from the balloon to the ground. It is captive to the mainline and the haulback. The butt-rigging attached to the eye of the mainline is brought within reach of the rigging crew by the haulback line. When the returning chokers have been uncoupled and the chokers which are secured to logs have been attached,

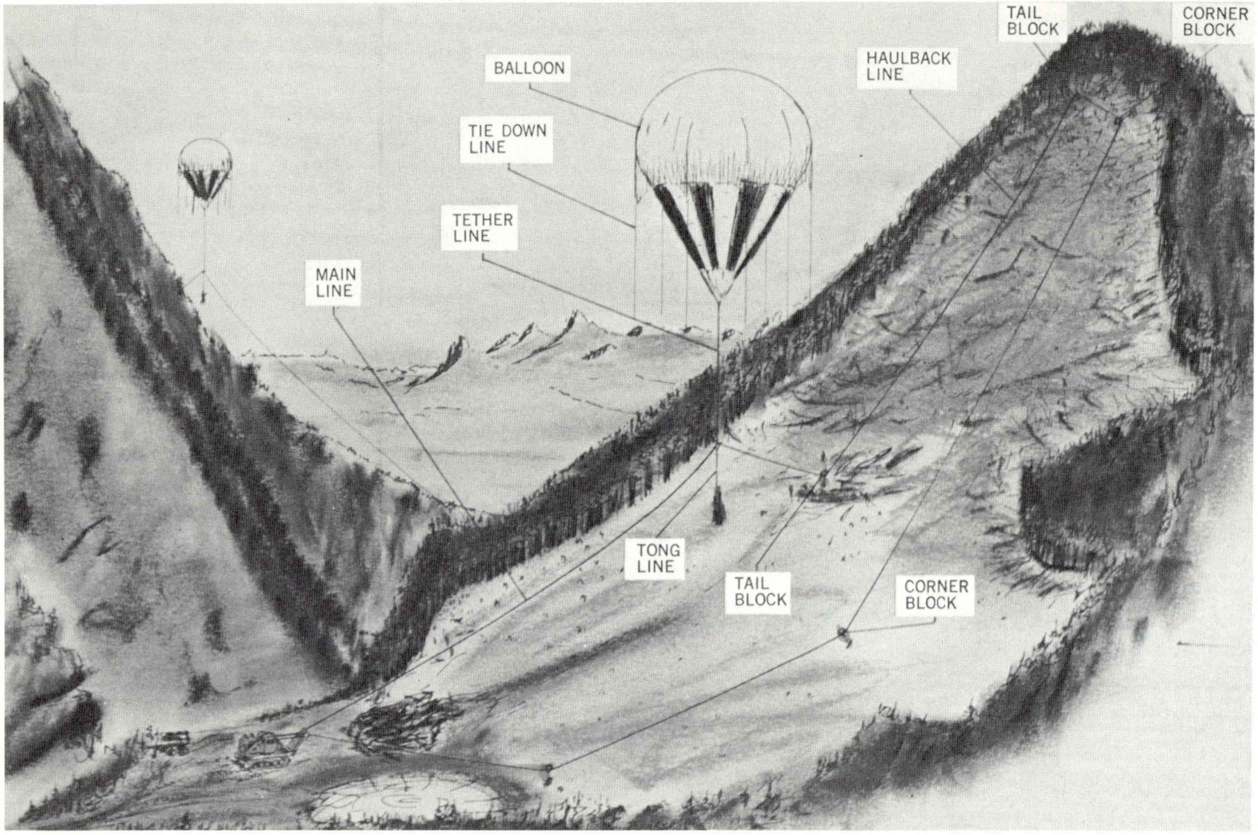

View showing the Washington Model 608 Aeroyarder working on a clear-cut operation down hill. This system is equally adapted to flat terrain. This is truly flying logs.

Courtesy Papé Cat, Eugene, Oregon.

Helicopter approaches landing with a fair-sized log. *Courtesy Loggers World, Chehalis, Washington.*

Helicopter lifts two logs to take to the landing where they will be loaded on a truck. *Courtesy Loggers World, Chehalis, Washington.*

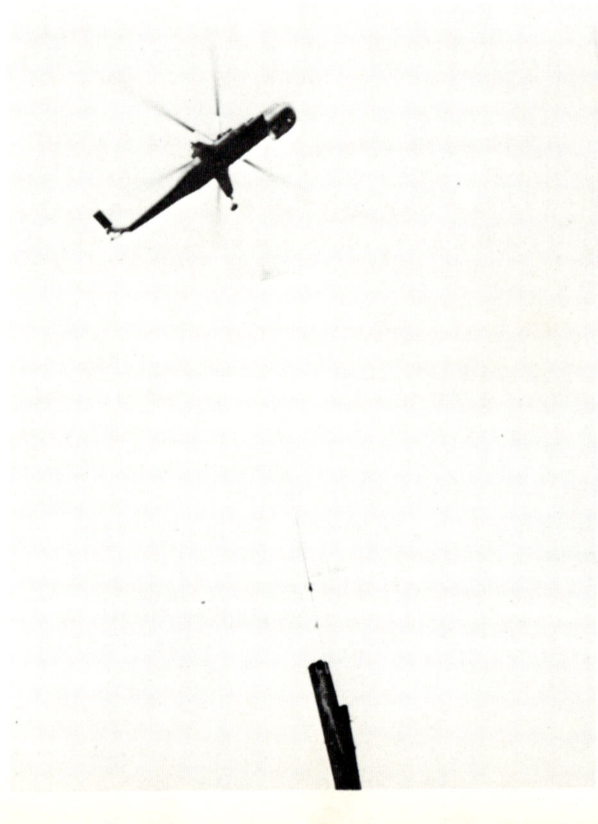

the balloon and its load are allowed to rise by slacking the haulback, and are pulled to the landing by the mainline. Elevation of the balloon is controlled by the proper braking of the haulback when the mainline is pulling, and braking of the mainline when the haulback is pulling. Another proposed method of controlling the balloon is the use of two lateral lines to the balloon that would permit the balloon to rise and lower only in the controlled arc governed by the adjustable lateral tether guys.

Other innovative systmes have come along too. For example, we inspected a show where a Washington Model 208 yarder working with a Washington 110-foot telescoping spar, was rigged to be used as a running skyline. That type of system can be rigged to use a radio-controlled slack-pulling carriage, or to use buttrigging and chokers. The hydraulically elevated tower, popular so few years ago, is already obsolete, and the cable-raised spar (or tower) is generally preferred. Use of these portable towers eliminates the need for properly located spar trees, as they are an integral part of the diesel yarder itself. Such spars, however, obviously still have to be guyed substantially. Another system, also Washington, is the Model 108 Infinite Ratio *Skylok* yarder using grapples instead of chokers.

An advantage of both of these latter mentioned systems is that they both use a running skyline which, in effect, operates very much like the slackline system. One principle has been common to all efficient logging operations since the turn of the century, that is the value of the upward pull on the logs (high lead) if not actually flying them to the landing.

Now, around every logging operation it will be noted that the men wear hard hats. Safety rules are stringently enforced and injuries are few, casualties rare.

A controversy has burned for years over the two alternatives, selective logging versus clear-cutting. Proponents of clear-cutting believe the chances of a good stand of second growth are better when everything is removed and the ground is somewhat stirred up in the process.

It is no longer lawful to leave the logged-off land littered with unwanted small logs, downed snags, and brush. The reasoning in the law is not just that the land looks better when the debris is removed, but that danger of fire is reduced. The net effect is that everything unfit for lumber must either be ground up and hauled out, or hauled out and ground up; both methods are used.

Ground wood can be made into softboard, such as Firtex. Chips can be used for attractive chipboard. The Bohemia Corporation has developed a number of uses for fir bark alone. Some of these have already been mentioned. Bohemia's Bark Products Plant went into operation in 1974. There are 80,000,000 pounds of fir bark available annually from the timber that is logged in the Northwest.

When you drive up into the timber on a logging road these days you do not pass a solid line of chunks, stumps, limbs, and trash where the blade has pushed them out of the way in grading the road. Stumps will still be seen but the debris has been removed. You can, if the logging company permits it, drive to the scene of the logging because most of it is done by trucks, and trucks must have roads.

The trucks you pass on the logging road will include some makes not mentioned before. Along with the KWs and Jimmies, Cornbinders and Macks, you will see Autocars and Peterbilts. The latter have been favored for logging in recent years.

You will not be greeted from a mile away by the tumult of two or three yarding, loading and swing donkeys grinding, roaring and steaming as they bark at the sky. When you get close enough you will hear the deep roar of diesel-powered yarders and loaders. In very few places will you find the 2500 to 3500-foot skylines. These were necessary in the days of railroad logging when roads could not readily be brought to the timber, but the timber had to be brought to the road, and it was urgent to reach out as far as possible from the railroad. If you come to a cutting area, you will hear the angry snarl of chain saws. Where the cry "timber" used to be heard for 1000 feet, even in the timber, now you are lucky to hear the most lusty yell 300 feet away. If the faller's power saw is not snarling, the saws of two or three buckers probably are.

At the landing there is not quite the desperation of movement by the men as in the old days. The new machines handle timber more unhurriedly and gently than the old ones used to. I must believe that no power on earth has ever moved logs from the tail tree to the railroad with the positive, irresistable speed of a big Skagit or Willamette two-speed skyline donkey.

Where one-log loads not only on trucks but on railroad cars used to be the commonplace, now they are the exception to say the least. When you follow a log truck down the road, count the logs. You might find from three to thirty on a load. I can't help wondering what use they are finding for the logs we used to call *dog's hair.*

Maybe logging methods have improved, the timber has not. Maybe the price of lumber, pulp or wood chips has made it profitable to log doghair. I know I'm glad I'm not driving one of the trucks that is

450-horsepower diesel engine tram works, Washington 208 yarder, telescoping tower, slack line with carriage. *Courtesy Papé Cat, Eugene, Oregon.*

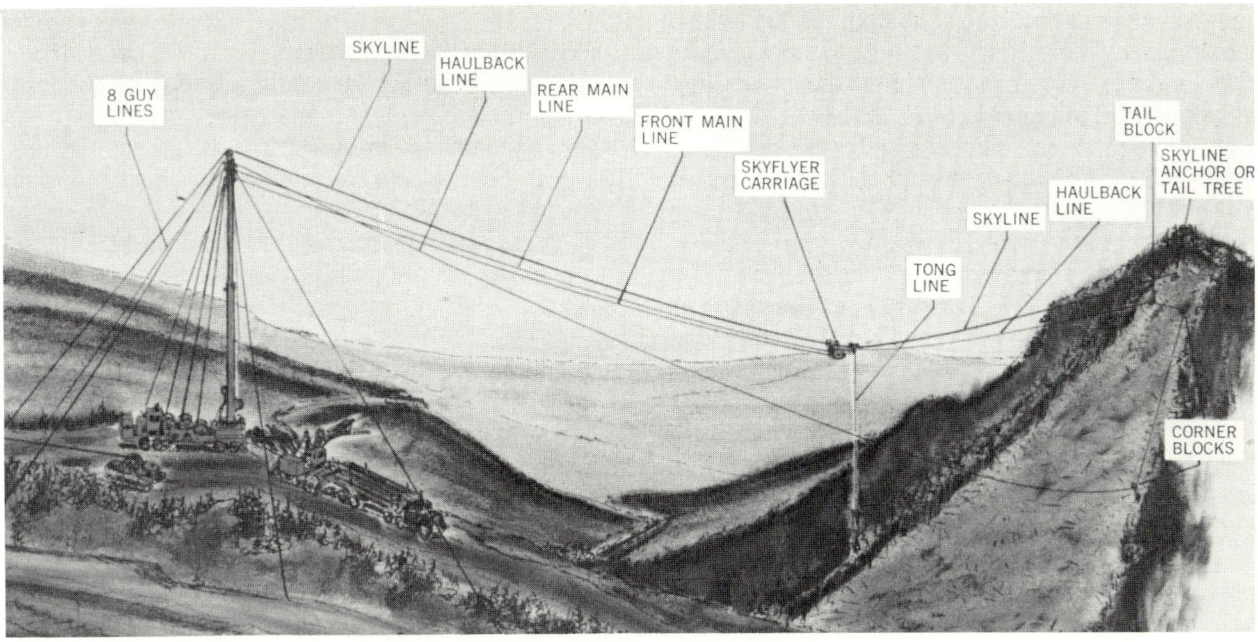

View showing Model 408 Skyflyer Yarder, Washington Skyflyer Carriage which is mechanically powered, and a Washington Self-propelled Skyline Drum Unit, clear cutting. This arrangement works equally well on up hill or down hill logging.

Drawing shows possible use of 408 Skyflier yarder. *Courtesy Papé Cat, Eugene, Oregon.*

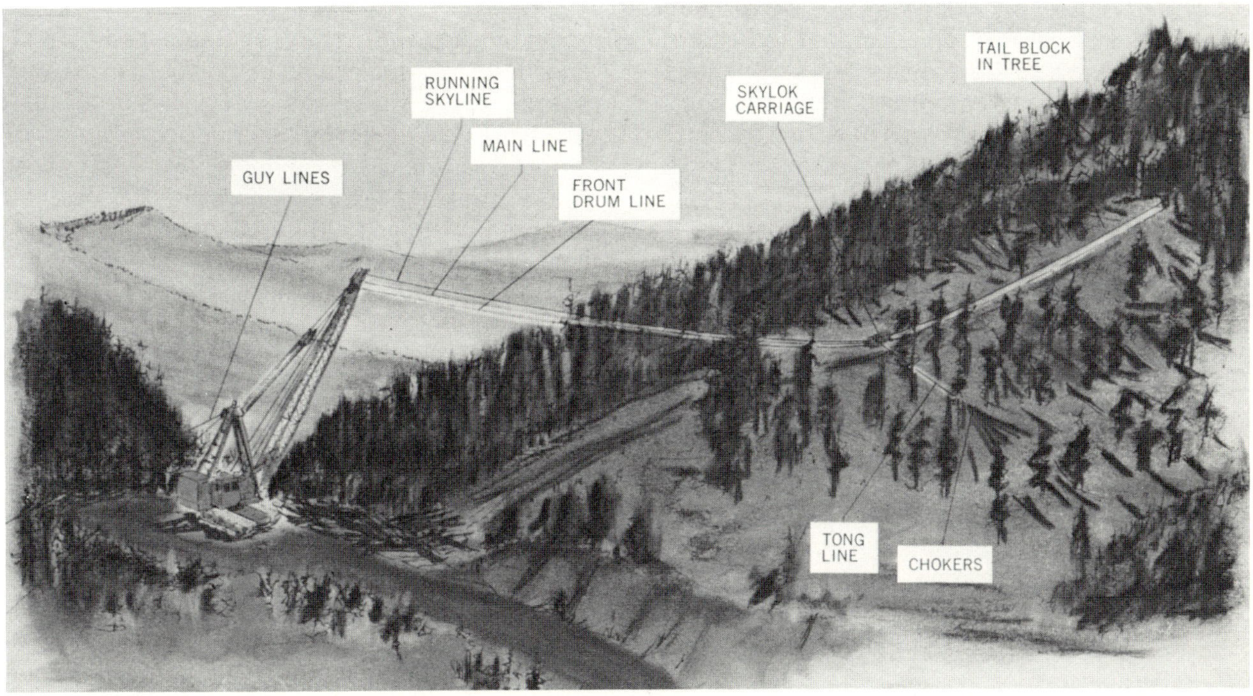

View showing a Washington Model 78 Infinite Ratio Skylok Yarder selective logging.

Courtesy Papé Cat, Eugene Oregon.

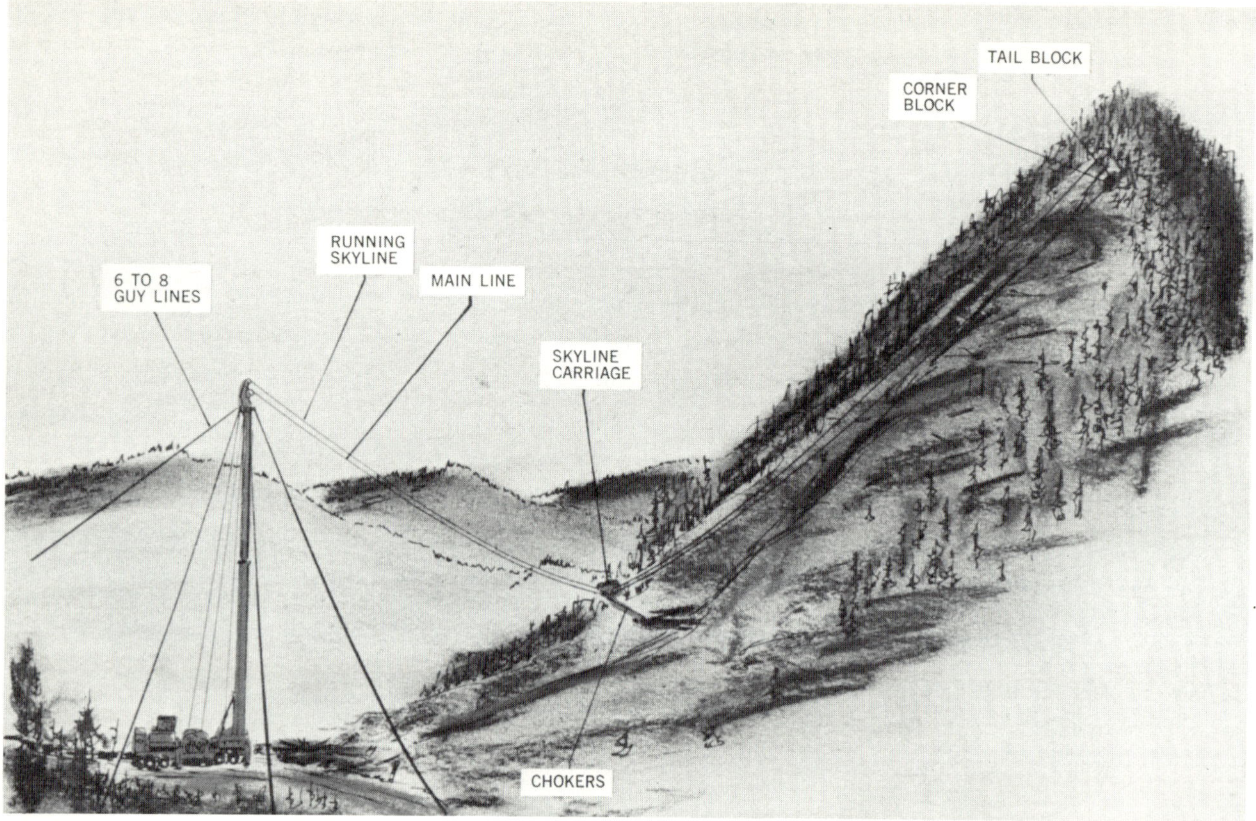

View showing a Washington Model 208 Yarder with a Washington 110 foot Telescoping Spar (Combination Model 208-110) rigged to be used as a running skyline. This system can be equipped with a slackpulling carriage, radio controlled, or with butt rigging and chokers.

Courtesy Papé Cat, Eugene Oregon.

supposed to pay for itself by hauling that kind of log.

Yes, the new ways may be better. I certainly know the old wood burners are no longer economically practical. But I'd give all I own to once again spend a month in one of the old logging camps. To watch the train come down the hill in the evening with a train load of logs that *were* logs. To watch a monster snubber let loads of logs down an incline so steep a man nearly had to climb it like a ladder. To stampede the cookhouse when the old triangle jangled its tune in the evening after a hard day's work. I even miss the old "briar" (bucking saw) so skillfully filed by the camp saw-filer that you could run it back and forth full length and watch it drop through a big log a quarter of an inch with each stroke. I even miss the tough old "hookers" (hooktenders) who used to bellow out a prolonged sentence of choice and skillfully selected profanity. That used to be referred to as "sayin' yer prayers". It was said of one hooker that he took off his hat and said his prayers, and all the little hemlocks bowed their heads.

I miss the evenings in the bunkhouses, where anything was liable to happen, and usually did. I wish I could hear again an old Lima Shay, or a Heisler heaving and chucking its way up to the woods with a string of empties.

Practically all the remaining old growth fir is in forest reserves or national parks. Logging companies up and down the Pacific Coast west of the Cascades are logging mostly second growth. A few of the old companies that "acquired" their timber over half a century ago still have some old growth left, and it's worth billions. The country I was familiar with as a boy and as a young man has all changed. The big timber is about gone, and has been replaced by second growth that is being taken out, not in respectable loads but in bundles.

It is a sad thing to see the vast, magnificent, virgin timber resource decimated. Trees that stood 500 years before Columbus set sail to try to prove the earth was round are gone now. They will never return. But, I am old-fashioned and sentimental, and as guilty as anyone else, I helped log them out.

110-foot hydraulically elevated tower, integral part of diesel yarder. *Courtesy Bohemia Corporation, Eugene, Oregon.*

GLOSSARY OF LOGGING TERMS

BACK-CUT: Cutting the rest of the way through the tree from the opposite side of the undercut.

BACK STRAPS: Lines from the rear of donkey to stumps to hold it in place when it is pulling.

BELL HOOK: Floating bell on a choker into which the knob-end or the nubbin can be pushed in endwise and turned into a slot.

BIGHT: Any spot within the curve of a "siwashed" haulback, dangerous when the haulback "saws" through the stump or slips over it.

BUCKING: Cutting the felled trees into most suitable lengths.

BUCKSKIN: A dead log that has lost its bark and turned a sun-tan color.

BULL-BLOCK: A heavy steel pulley with bearings through which the mainline passes for "high-lead" logging.

BULLBUCK: Foreman over the buckers — often the scaler also.

BULL COOK: General roustabout handled food stocks for the cook, built fires in the bunkhouses, etc.

BUNKHOUSE: Where men slept, read, played cards, told tall tales, and rarely fought.

BUTT-RIGGING: Heavy steel assembly at the end of the mainline; contains slots into which choker knobs can be inserted for the pull; contains a mainline swivel toward the donkey, and a swivel on the opposite end for haulback.

CAULK SHOES: (called "cork" shoes) Hard-soled shoes made of the best leather with soles so hard they don't bend and will hold spikes to keep a rigging man from slipping off logs.

CHASER: Man who uncouples the chokers at the landing for their return to the woods.

CHOKER-SETTER: (Rigging Man,) Man who buckles the cable "choker" around a log, and secured the other end to the "butt-rigging".

CLIMBER: Similar to high-rigger, but is not generally the one who plans trees to be used for spar trees.

COOKHOUSE: Not only where the food was cooked for the crew, but also where it was eaten.

CROTCH LINE: Two lines that are clevised to one, below loading block on a guy-line to bring the log over the load when it is lifted.

CRUMMY: The railroad car, or enclosed truck, used to transport the crew to and from work (in early days might mean infested with bedbugs).

DONKEY: Steam, gas or diesel-powered machine (usually on a large sled) for pulling logs.

DONKEY SETTING: Leveled off place large enough for donkey and any other needs.

END-HOOKS: #1, huge L-shaped hooks to put around end of the last of several logs when pulled on a skid road by a big donkey called a roader; #2, cupped hooks with spikes in the center, used on soft pine to prevent tearing out large chunks and possibly killing somebody if tongs had been used.

FAIRLEAD: A system of four concave, conjoined rollers to keep the lines always going directly to the drum, no matter what the direction it may go beyond the fairlead.

FALLING: The felling of trees, always called 'falling'.

GANDY DANCER: Tie tamper (or setter) on the railroad.

GUY LINES: Cables that go from the spar tree top and from its center to surrounding stumps to prevent tree from wobbling when donkey is under heavy strain.

HAULBACK: Smaller line used to pull the mainline back to the "woods" for its next turn of logs.

HAULBACK BLOCK: Tail-block.

HAYRACK BOOM: Straddles the tree because it is made of two logs planked together. One tong is hung at the front end of one of the hayrack logs, the other is strung near the rear of the opposite log of the boom. With this angle on the log being lifted when the boom is swung over the load the log will be in line with the load.

HAYWIRE: Idiom for guinea line, small enough to be strung by hand, thus enabling the other lines to be strung by power from the donkey.

HEEL BOOM: Uses only one tong put on the log short of center. When lifted, the log will heel on a specially built pad (planks or steel), then additional lift brings the entire log up to level; it is then swung over the load and placed in position.

HIGH LEAD: When the mainline passes through a "bull-block" hung high in the tree to give lift to the pull to help get logs over stumps and other obstacles.

HIGH RIGGER: Man who climbs, limbs and rigs tree for logging. Also usually selects trees that are to become spar trees.

HOG: Old term for steam locomotive. (In a mill the machine that chops hog-fuel used to fire mill boilers, preferable to sawdust.)

JAGGER: Broken cable stand capable of inflicting serious and painful wounds.

JILPOKE: Limb or slab placed diagonally between a log and a stump to change the course of the log long enough to get it by the stump when pulled.

MAUL: Steel hammer of six to ten pounds for driving wedges into back-cut. Also splitting maul has a splitting blade on one side, hammer on the other.

MEAT WAGON: Ambulance for the injured, or dead.

MOLLY HOGAN: Strand of cable containing about five wires. When these are wound into themselves circularly, because of their kinky wavy construction they are almost certain never to pull out under heavy lifts or pulls.

PARBUCKLE: Bight of line placed around a log causing it to roll when one end of the line is pulled by the donkey — or any source of power.

PASS-LINE: Guinea line used for pulling blocks and lines up to the high rigger.

PETER HOOK: (also called C. hook) Hook on the bitter end of a choker, shaped like a thick C.

POWDER MONKEY: Man who handles explosives.

PUSH: Foreman. Also called **the man, the rod.**

RIGGING GLOVES: Heavy gloves made to fit perfectly, of the best leather to protect logger's hands from "jaggers". The old standby brand were Churchill gloves until that manufacturer was boycotted by loggers because it was a non-union company.

RIGGING SLINGER: Lead-man in charge of the choker setting operation.

RIG-UP GANG: Crew who assists the high rigger and move donkeys.

SCALING: Measuring logs to mathematically determine board-foot content.

SCHOOLMARM: A forked tree, usually must be felled one fork at a time.

SHEAVE: Best described as the wheel in a pulley though this is no woods term.

SIWASH: When a haulback is allowed to be diverted from a straight line by being on the wrong side of a stump or tree.

SKYLINE: Two-and-quarter inch line (generally) stretched taut between two trees to keep a constant lift on the turn of logs to prevent hang-ups.

SLACK-LINE: Same as skyline except it is slacked between turns and tightened by a special drum on the donkey when the turn is ready to be pulled.

SNAG: Dead tree generally without limbs, often with top broken off, of little use for lumber.

SQUAW-HITCH: Bight of line put around half or more of the end when a log is partly buried.

SQUIRREL: Half a log used as a counterweight to return a hayrack boom to position again after it has been pulled by donkey power the other way.

SOUP: Superintendent. Also nitroglycerin.

STEEL GANG: Separate crew who laid the steel rails on the railroad.

SWING CHASER: This man must do the same functions as chaser, but also couples the chokers to the butt-rigging of the skyline donkey or swing donkey.

SWING DONKEY: #1, skyline donkey (almost always two-speed). #2, rarely are three trees and three donkeys used. When they are, the donkey that swings, turns from the yarder tree to the skyline donkey's tail tree, also be skyline or slack-line, is called the swing donkey.

TAIL-HOLD: Tree and cable strap to hang the haulback block.

THE TREE: Not just a tree, but **the** tree, limbed and rigged for a "spar tree".

TIE-DOWN LINES: Cables across the front of donkey to two stumps to hold the front-end down against the upward impetus of the mainline on heavy pulls.

TIGHTLINE: The act of pulling on the mainline and the haulback both at the same time. Used to give additional lift to get a log up and over a stump. If a logger inadvertently straddles a line and the line lifts sharply, he may be tightlined twenty feet in the air and rendered remarkably sterile instantly.

TIN PANTS: Pants made of heavy, tightly-woven, hard material, then treated with paraffin for waterproofing, but more importantly to create a surface that will resist the dangers of jaggers or sharp pieces of limbs. It has been said they will stand by themselves.

TONGS: Similar to ice tongs, big enough to span a good-sized log for lifting it to load.

TURN OF LOGS: The number and size of logs required to make up a load for yarding.

UMBRELLA: Fully-rigged tree with all guy lines and other lines affixed.

UNDERCUT: Face cut in a tree on the side of the direction of intended fall. Properly done, in conjunction with the back-cut and wedges, the tree can be dropped with impressive precision.

UNDERCUTTER: Device to attach to a log with a roller for the back edge of the bucking saw to ride upon, when the log must be at least partially bucked from the bottom upward.

WEDGE: A falling wedge is broad, thin, gently tapered for holding tree from settling back on the falling saw. Also to help direct its direction of fall. A splitting wedge is narrower and tapers rapidly; driven into block to split.

WHISTLE PUNK: Pulls whistle to signal donkey puncher. Term, **Punk** (Shakespearean): Most lowly job in the woods.

WIDOW-MAKER: Hanging limb, broken by another falling tree, can fall and kill.

WOODS: #1, place where rigging crew are working. #2, entire area being "logged".

YARDING: Pulling of the turns of logs to the first (or yarder) tree.

INDEX